D0532305

fAce the human

the human fACe

BRIAN BATES WITH JOHN CLEESE

London, New York, Sydney, Delhi, Paris,
Munich, and Johannesburg

Publisher: Sean Moore
Editorial Director: Chuck Wills
Project Editor: Barbara Minton
Art Director: Dirk Kaufman
Production Director: David Proffit

0-7894-7836-6

First US edition published by
Dorling Kindersley Publishing, Inc.
95 Madison Avenue
New York, New York 10016

This book is published to
accompany the television series
The Human Face,
first broadcast on BBC1 in 2001
Executive Producer: Nick Rossiter
Series Producer: Sally George

Published by BBC Worldwide Limited,
80 Wood Lane, London W12 0TT

First published in 2001
Copyright © Brian Bates and
John Cleese 2001
The moral right of the authors has
been asserted

All rights reserved. No part of this
book may be reproduced in any form or by
any means without permission in writing from
the publisher, except by a reviewer who may
quote brief passages in a review.

Commissioning Editor: Joanne Osborn
Project Editor: Sarah Lavelle
Copy Editor: Christine King
Art Director: Linda Blakemore
Designer: Bobby Birchall, DW Design, London
Picture Researchers: Miriam Hyman,
Rachel Jordan, Frances Topp, Frances Abraham,
Sophie Hartley

Set in Garamond and Gill Sans
Printed and bound in the UK by
Butler & Tanner Limited, Frome
Color separations by Radstock Reproductions
Limited, Midsomer Norton
Jacket printed by Lawrence Allen Limited,
Weston-super-Mare

contents

introduction

W E HAVE BEEN ENCHANTED BY FACES FOR A VERY LONG TIME. Nearly three million years ago, a cobble-sized stone with the striking markings of a face – bulbous eyes and a dramatic gash of a mouth – was picked up by one of our far-distant ancestors and carried back to the cave. Called the Makapansgat cobble after the province in South Africa where it originated, it was discovered by archaeologists in the early twentieth century still in the cave, along with skeletal remains of its early owners. Microscopic and geologic examinations by archaeologist Robert Bednarik in 1997 revealed that the cobble had been carried a distance of at least 32 kilometres to the cave. They also showed that the cobble had acquired its very distinctive human face marking through natural geologic activity, rather than any carving or cutting. There is no obvious reason for the cobble to have been carried from its original position and stored in the cave, other than the proto-human's fascination with its facial features.

And we are drawn to faces even as tiny babies. American and British psychological studies have shown that nine minutes after being born, when we can barely focus our eyes, we prefer to gaze at faces more than any other object. Babies look almost as long at the eyes on a face as at the whole face – the ever-shifting movements of our eyes capture the infant's attention. We come into the world primed to connect with the faces around us. Our fascination with faces is inborn, and continues all through life.

Our brain recognizes a montage of features as a face, in a process specialized in the brain and different from seeing anything else. Our face is not just an object, but is something that we are born to perceive uniquely. Children's drawings of people perch enormous heads on stick bodies – but they reflect accurately the true significance of the face, for the features do seem to be central to our understanding of our identity.

In fact, the face is such an intimate part of our lives that understanding its origins, how it works and what it means, is a way towards defining who we are. This book explores the human face as a journey into ourselves.

However, our face is so familiar to us that to understand its mysteries we need to suspend our preconceptions and see it as if for the first time. In this book, to rediscover our faces anew we ask such questions as, why do we have a face? Where does it come from? We explore first the evolution of our faces, from the feeding holes of prehistoric worms floating in primeval oceans, to the complex, familiar structure we have now as primates. We inquire how we became so dependent on our visual senses, and how that helped to shape the evolution of our features.

And those features are unique in every person. For each of the six billion individuals on Earth, the face is an identity tag. How did this staggering variety arise? Why are there distinctions between the sexes? How can we remember the different faces we see? And, we ask, what happens if you are one of the small number of people who cannot even register the features of a face in the first place? Once we have recognized a face, and assigned it an identity, how can we interpret its expressions – how do we 'read' it? Expressions communicate our emotions faster, more subtly and more effectively than words ever do. We use them to reveal our feelings – and to hide them, as well. We explore the difference between the fake and the genuine, and their role in our complex social lives.

What if the face making the expressions is beautiful? Do we respond to it in a different way? We consider what we mean by 'beauty'. Are judgements of beauty purely a matter of taste, or are some features universally admired? Physical beauty and sexual attractiveness often go together – so are they biologically based? Are we programmed to respond to certain faces? Then again, should not beauty be more than skin deep? Surely a superficial view of beauty leads to vanity. We explore this 'deadly sin' and its relevance for understanding the meaning of the human face. Is it so deadly? Perhaps it is just shallow and frivolous (although worth billions to the cosmetics industry)? But a closer look into the mirror of vanity reveals issues of peer pressure, conformity to norms – and a concomitant intolerance of 'difference'. A darker side of the human psyche.

Our exploration of the human face finally takes us to a realm of hype and super-hype: the phenomenon of fame. Of the billions of unique faces in the world, a few become famous. Their images inundate our daily lives. From television screens and magazine racks, advertising billboards and movie posters they stare out at us, clamouring for our attention. But this is a two-way process; we are endlessly intrigued by the faces of the famous. Why are we so fascinated by fame? What is it about these people that persuades us to give them such power, money and influence? It is partly to do with commercial interests, of course – famous faces sell products. But beneath this surface, other forces are at work, deeper forces stretching back to our tribal past, and a spiritual dimension of which we are now unaware.

This exploration of the nature and meaning of the human face – a fresh and revealing look at that most familiar part of ourselves – is a journey of self-discovery. It takes us from the faceless world of primordial origins to the spiritual realms of the gods.

origins

CHAPTER 1

Below Five hundred million years ago, most of the Earth was under water. Early organisms looked something like this brown blubber jellyfish and sea star, and they did not have faces.

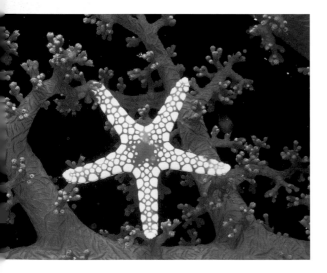

THE FACE YOU SEE IN THE MIRROR HAD ITS BEGINNINGS IN THE PRIMEVAL SLIME – AT THE BOTTOM OF THE SEA. Although we take our faces for granted, they are a relatively recent invention. For most of the planet's history, creatures have been faceless. Tracing the journey from them to us is, like most evolutionary stories, full of staggering timescales and unpronounceable names of long-extinct creatures. But they are all 'part of us' today, for their features were the forerunners of our features.

The Earth was born 4.5 billion years ago. This timespan is beyond our comprehension, so if we imagine the time between then and now condensed into one year, it becomes a little easier. Life on Earth did not begin until Valentine's Day, 14 February. Then, for nearly 4 billion years – that's ten months of this condensed year – the Earth was inhabited only by bacteria. At this point in time, things began to change.

The world looked very different back then, 580 million years ago. Most of it was covered with water. Under the waves (the moon has been pulling the oceans for a very long time!) floated a variety of faceless bags of tissue without shape or bone structure. In this primordial soup, all animals were like food-gathering sacks. Nutrients in the seawater sloshed in and out of them. They had no motor nerves so were unable to move about voluntarily. They collided with food and took it in almost by accident. From fossils, we know they looked a lot like modern-day sponges, jellyfish and sea squirts. Ranging from just 1 centimetre to 1 metre across, these animals drifted on the currents, with no faces to tell in what direction they were headed.

Fishy features

On 15 November of our evolutionary year, there emerged a creature called Pikaia. It lived on the seabed 520 million years ago. Pikaia was rather unusual because, unlike any of its fellow marine animals, it had a hole at one end. It looked like the empty plastic case of a biro pen. The simple opening was a primitive mouth that was able to inhale water and sift out food particles. This was a more efficient feeding method than ever before. The animal was less passive than its ancestors; it could suck in food, and that meant it dined in relative style.

Pikaia might seem harmless and insignificant to us but, to the bacteria it fed upon, that gaping mouth would have looked pretty alarming – if only they had had eyes to see it. But eyes had not developed yet. The face – as we know it – began with a mouth. And Pikaia, this worm with a hole at the end, is our most venerable facial ancestor.

By late November in the evolutionary year, Pikaia's great, great, great, great (and many more) granddaughter had been born. She was a Conodont (we are not sure if this particular Conodont was a 'she', but let's alternate gender, rather than rudely calling our ancestors 'it'). Just 5 centimetres long, she looked rather like a small eel. And she might have been able to see Pikaia – if only Pikaia had not become extinct – for she had two areas of light-sensitive cells around her mouth to help guide her towards her prey. So the Conodont was the first creature to have two 'eyes'. Her eyeballs were set in a

Above The hammerhead shark has eyes at either end of a wide face. All sorts of arrangements of the features have arisen during the course of evolution.

ring of cartilage, with a single lens and a pupil. These primitive, bulging eyes permitted her to spot darker, more densely populated areas in the water that indicated profitable feeding areas. She was able to propel herself towards them, and once she got there her tongue's fifteen primitive teeth made short work of her prey.

Once these sorts of creatures could move through the water, they explored more varied habitats. They began to require increasing amounts of information about their new surroundings in order to detect their food. Rudimentary sense organs started to emerge to help them to smell and hear their prey, as well as to avoid predators. They developed primitive ears – areas up by the eyes and mouth, sensitive to changes of pressure in the water. This means that by this stage in evolution there are fish that can eat, see, bite, smell and hear.

Then came a nose. On 25 November of our year, the Sacambapsis evolved. He was about 10 centimetres long and covered in protective scales, like a fish. He had a mouth like a letterbox, and on the upper edge of this mouth he had two tiny breathing gills. He was able to smell food in the water and could continue to breathe while he ate.

Next came a jaw. This major step forward came 430 million years ago. It sounds unexciting, but until this point these aquatic creatures had been unable to close their mouths, or stop food escaping. The Acanthodian was a fish with a moving jaw. She could bite, and her lifestyle was very different from that of our earlier ancestors. Instead of being a relatively passive filter feeder, waiting for food to seep into the mouth and hoping to digest it before it seeped out again, the Acanthodian was an active predator. She chased after food and then dissected it with precise efficiency. Except that on occasion she was too hungry to bother with the dissecting bit; one fossil Acanthodian was found with an entire smaller fish in its stomach.

This fishy story of our ancient evolution seems so vastly long ago that we might wonder how it is relevant to the human face today. But remarkably, our former existence in the water is still present in gill slits that form at the sides of our neck in the womb, and which then become sealed over before we are born.

So in this way, gradually, our major facial organs ended up in their familiar place. Along with the mouth came the other facial features that help

Opposite The red-eyed tree frog retains the bulging eyes of the conodont, so useful for locating food in aquatic habitats.

obtain food: a nose, for smelling where it is and making sure it is fresh; eyes, for seeing prey; and then various refinements like jaws and teeth for crunching up our food.

It was all done through natural selection – it was the organism with the face most efficient at taking in food that survived over its competitors. But if any of the strange sense organs on the creatures swimming around the sea 520 million years ago had been more successful in evolutionary terms than the ones we have described, the human face might have looked quite different.

The story also makes us realize what a face is. Whatever significances it holds for us today, the story of the human face reminds us that it evolved essentially to make us more efficient eating organisms. Enjoying our food seems the least we can do, after all that evolutionary effort.

Dinosaur teeth

Then the tale of our face took a slightly more recognizable step. We headed out of the water. On 2 December in our evolutionary year, a little reptile called Acanthostega stuck his nose above the water and peered over at the land. Getting on for a metre long, facially this creature looked like a fish. But he was equipped with stout limbs and could walk on land. Even more importantly, he had tiny nostrils that could breathe air. This opened up a whole world of evolutionary opportunity. There is far more oxygen in air than in water, and it is therefore easier to obtain energy from the air. So our ancestors quite sensibly clambered onto dry land.

Over eons, various of the land-based reptiles became larger and more powerful. Some dinosaurs were enormous. But their jaws were rigid and rather inflexible. That makes them rather disappointing in a couple of ways. Firstly, they could hardly have been very expressive. Looking at a dinosaur face, it would have been hard to tell if the creature was sorrowful. Or angry. Reflective? Ironic? Hungry? We will come to the wonders of flexible facial expressions in a later chapter. And what's more, the dinosaur's teeth were rather boringly all the same size and shape, which meant that he was simply able to bite and swallow.

This was a more significant limitation than might first appear, pointed up by the evolution of the Dimetrodon. This large creature, which looked like a dinosaur but wasn't, had a massive jaw with two different sorts of teeth, one set for biting and another for tearing food. She could manage a very primitive sort of chewing. That may not sound much, but it is one of the critical stages in the development of our faces – for the Dimetrodon, with a more flexible mouth, eventually evolved into mammals, and us – while the dinosaurs evolved into birds.

Below The Megazostrodon was our first mammalian ancestor. It looked like a large shrew.

Warm blood and whiskers

The arrival of the mammals, on 17 December of our evolutionary year, was a key advance in the evolution of the human face. A creature called Megazostrodon was our first mammalian ancestor, with the 'warm-blooded' ability to keep his body temperature constant. He looked a bit like a large shrew or small opossum, with a long pointy snout, big eyes, a hairy face and external ears. The large ears acted as trumpets, concentrating and locating sounds with accuracy – crucial for any animal that lives mainly on the ground. His long twitchy nose was equipped with whiskers and could follow scent trails – perfect for a creature that was largely nocturnal. Importantly, Megazostrodon had a much more flexible jaw than other creatures, which meant he could nibble and chew food more effectively.

As a mammal, he had all sorts of characteristics that affected the appearance of his face. While reptiles cannot internally regulate their body temperature, and need a hard skin in order to stop the ambient temperature of the environment frying or freezing them, mammals' warm-bloodedness means that they have thinner

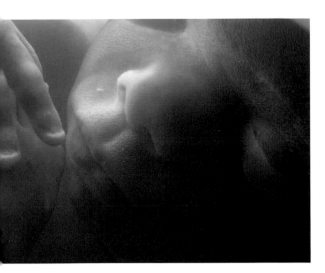

skin, which is insulated by hair. Hair is made of the same protein that forms scales in reptiles – keratin – but is much more efficient at regulating body heat. The skin beneath is flexible and elastic.

Mammals give birth to live young, who are suckled in the first months of life. The baby needs to be able to form a pout with its mouth to secure nourishment from its mother – and this is enabled by the strong muscles and flexible skin around the mouth. Mammals later started to use these facial movements as a means of communication. This stage of evolution could be said to mark the birth of facial expressions.

But despite all these illustrious qualifications, and though he had better table manners than his reptilian ancestors, and big ears to help protect him from danger, sadly the Megazostrodon became extinct.

Swinging from the vines

However, some of our shrew-like ancestors continued to adapt and made the most of the food opportunities available. They started feeding during the day, and also headed into the trees. They grew bigger and stronger. And 25 December of our evolutionary year saw the arrival of the first primates. Called Aegyptopithecus, she looked not dissimilar to modern-day lemurs. Her face had changed almost beyond recognition from that of her ancestors. Her eyes had rotated towards the front of the head, giving two overlapping fields of vision – essential for judging distance accurately in the trees. Her face was flat and the jaw had retreated beneath the brain. This was because of a change in diet. Our mammal ancestors had eaten mainly insects and the jaw was used largely as a scoop. Primates ate leaves and fruit, and needed to grind these down into pulp in order to digest them. Grinding increases in efficiency as the jaw shortens.

Aegyptopithecus was the size of a small cat, and appears to have been the most sophisticated and intelligent creature on Earth at the time. She lived in the tropical forests of North Africa and ran along tree branches, collecting fruit and leaves. She is the ancestor of the great apes, chimps and humans.

The story is now beginning to become very familiar. As apes developed,

Above Human and other mammals' babies need to be able to form a pout with their mouth to suckle nourishment from the mother.

Opposite Humans and chimps are close relatives – but the human face developed a distinctive shape with a larger skull, and a prominent nose and chin.

the skull got larger and began to assume a distinctly human configuration. In apes and early humans the enormous jaws, and reinforcement over the brow to support the jaw, show that the most important facial activity was still biting and chewing.

Today our DNA is 98 per cent the same as a chimp, but this close association leaves room for many important differences. Over the course of evolution the human face developed a distinctive form. One facial change between ape and human has been the diminution of the canine and incisor teeth. In apes the teeth have long roots and the canines are enormous, for stabbing and cutting food. Human teeth are much more shallowly rooted and fairly proportionate in size, since we cook and/or break up our food before we put it in our mouths.

As the teeth shrank the jaw became smaller and was pushed backwards. The reason that we have chins is that the lower teeth row shifted backwards more quickly than the upper, leaving a band of jawbone behind. Over the 6 million years since then, our skull expanded to accommodate an enlarging brain; and our hair receded, allowing better regulation of body temperature and giving us increased expressiveness. We acquired a small, protuberant nose. The eyes and nose moved closer together. If today we could meet 'ourselves' from this period of evolution, we would be looking at a face that was becoming eerily familiar.

Lucy in the Sky with Diamonds

Perhaps the most famous of our primate ancestors from this time is known as 'Lucy'. She was discovered by palaeoanthropologists working in Ethiopia, who named her after the Beatles song they were listening to when they discovered her skeleton – 'Lucy in the Sky with Diamonds'.

Lucy lived at 6.24 a.m. on 31 December of the evolutionary year. She had a heavy face with projecting jaws. She walked upright, so only a small area of the body – the top of the head – was exposed to direct sunlight. This part of the head has retained its protective covering of hair even up to today.

Our brains continued to grow, and the pressures this put on the brain

case forced the chin and nose into a more vertical alignment, giving us the distinctive human profile which is shared by no other animal. The top half of our faces lost almost all their hair, except for tiny outcrops like the eyebrows, and this may in part have been due to the urge to communicate. Facial expressions, which convey information to and from members of our

Above An artistic reconstruction of Lucy, the most famous of our primate ancestors, who lived in Ethiopia 3.2 million years ago.

23

Above Portrait of Eve by Renaissance artist Lucas Cranach. The real Eve evolved in the hot deserts of Africa, and so needed protection from the burning hot sun. She would have had a lot of melanin – and black skin.

community, are much easier to discern and interpret on a naked face.

The smile, in particular, is visible from a very long distance. It was vitally important for us to evolve such an expression. As a hunting species, in primeval times, we had to become increasingly co-operative if we were to survive. The smile is a fast and effective way of sending friendly signals to one another. Other primates have friendly signals, such as lip smacking and teeth chattering, but, while effective at close range, those do not work so well at a distance.

And then, at 11.37 p.m. on New Year's Eve, *Homo sapiens*, the human being, appeared. In historical time this is 200,000 years ago. It sounds like eons to us, but is a mere sliver of time in evolutionary history. This event is celebrated in mythology around the world with a myriad creation stories. In the Christian biblical tradition, for example, these first *Homo sapiens* were the real Adam and Eve. The underlying structures of their faces and bodies were identical to ours.

Without knowing recent developments in fossil history, western artists used to imagine Eve as European, a fair-skinned blonde-haired beauty, standing in a lush green meadow. But the pale skin, hair and eyes of Europeans probably appeared only about 40,000 years ago, and perhaps even more recently. The first people more likely had black hair, as well as dark eyes and skin, and flourished in the dry, sandy savannah of the African deserts.

A tour of the features

The evolution of the face is basically the story of the mouth – from a small hole in the end of a floating organism, all the way to a jaw structure with teeth that can chew. We have covered the development of the mouth in explaining how our face came to be the way it is today. But what of some of the other features?

From top to bottom Nature
has experimented with various
arrangements of facial features across
all types of animals.

How does the baby's face develop?

The face develops rapidly — faster than any other part of the body. At just four weeks after conception, the embryo weighs 1 gram (0.035 ounces) and is only about 6 millimetres (0.25 inches) long. However, it is already beginning to reflect the stages of millions of years of evolution that have led to the present human form. The embryo resembles a fish or reptile embryo. It has the beginnings of a mouth and dark spots where its eyes will be. And at five weeks, near the head there are six cartilages on either side, symmetrically arranged and with appropriate muscles and blood vessels to form the gill slits of a fish. But then the embryo moves to later stages of our evolution.

A round head shape develops. The eyes, which started off on the side of the head, with the nostrils widely separated, gradually move round and forwards to cluster on the front of the head. By three months the baby has a fully developed tiny face and can swallow, frown and suck its thumb.

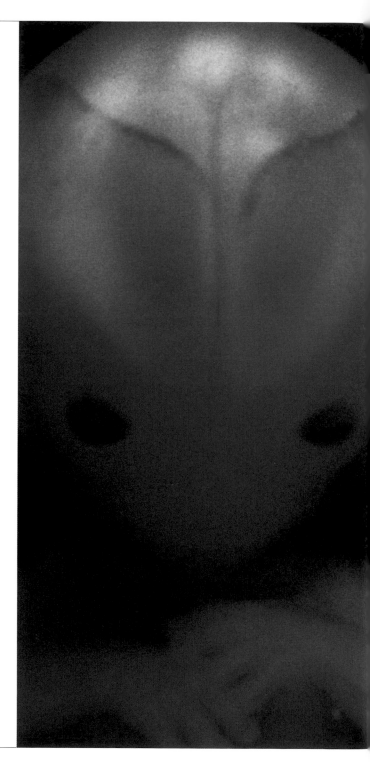

SKIN

Over all our features lies our skin, a remarkable, flexible, thin covering that protects us from the hazards of the world outside. It is water- and bacteria-proof, and if it is dark it is also sunproof. Because *Homo sapiens* evolved in the sunny deserts of Africa (hominids have their biological home in the eastern and southern third of Africa, especially Tanzania and Kenya), they needed to protect themselves from burning, and this is why they almost certainly had

black skin. Melanin is a form of colourant. It lies in the bottom layers of the skin. A lot of melanin makes our skin dark; a little, and our skin is lighter. Melanin acts as nature's own sunscreen, stopping UV rays from penetrating the skin and burning it.

Consequently, white skin has little protection against the sun. Skin cancer is caused by UV damage, which breaks down cell walls and helps tumours to develop. In the fifty years since sun bathing became fashionable, the incidence of skin cancer among light-skinned people has increased

Above Skin colours and types vary all over the world.

enormously. White Australians have the highest rates of skin cancer in the world, black Aborigines the lowest.

As humans began to migrate out of Africa 100,000 years ago they encountered new environments – places that were colder and damper than the ones they had known. Gradually, over thousands of years genes were selected which altered the colour of their skin. Given the prominent role skin colour plays in racial prejudice today, it is surprising just how recent are the Caucasian adaptations of pale skin from the formerly dark complexion.

Those of our ancestors who migrated to the north suddenly found that they were suffering unexpected problems due to their dark skin. Enough sunlight must penetrate the skin to set off a chain of chemical reactions that strengthens our skeleton. UV light activates steroids that synthesize Vitamin D, which then enables calcium to be laid down in the bones. Human remains from Sweden from between 17,000 and 10,000 years ago show inadequate calcium deposits. Even this recently in our evolutionary history, the skin of these ancient Scandinavians seemed to be too dark to allow the weak sunlight through.

Skin is like a sieve.

The density of the

mesh is altered by

melanin, which

controls the amount

of light permitted to

filter through.

Right The surface of the skin, under high magnification.

HAIR

It is likely that our common ancestors had black curly hair. These tight corkscrews allowed heat to evaporate from the head quickly.

The colour of the hair is once again determined by the amount of melanin. A little melanin results in blonde hair, and a lot, black. But redheads appear to have a mutation in their melanin cells that has changed their shape. While other hair colours are caused by light reflecting off sausage-shaped melanin molecules, redheads' melanin molecules are spherical and so the colour appears different.

EYES

Contrary to popular belief, the different colours of eyes are not caused by different colourants in the iris. Again it is all down to the amount of the brown colourant, melanin. Generally, people who have little melanin in their skin also have little in their eyes, causing them to appear blue. The more melanin you have in your iris, the darker your eyes look.

Left to right Young girl in Burkina Faso; blond European boy; elderly man in India.

Above Different coloured irises are due to the amount of melanin, which is a brown colourant.

Far right Diagram of a rod cell from the retina.

The evolutionary process whereby melanin was lost from the skin also caused eyes to become paler. Blue eyes are the rarest colour, because the gene that determines them is recessive and easily overruled by the genes for brown or green eyes. Blue eyes are widely admired, and this may be because their light colour make the size of the pupil very clear. If the pupil enlarges in response to desire or fear, it is more obvious in light-eyed people.

Our eyelid is composed of skin only one millimetre thick. It is slightly translucent, so that light can pass through it. We close our eyelids to blink, an action that stems from our aquatic origins hundreds of millions of years ago. When we lived in water our eyes were kept moist, but in air we need to protect our eyes from drying out. Each blink covers the eyeball with mildly

antiseptic tears that reduce the risk of infection to the eye. These tears also carry oxygen to the cornea of the eye.

NOSE

The shape of our noses seems to have evolved in response to environmental factors. Because the lungs need air at about 35 degrees centigrade, or roughly equivalent to our body temperature, and at 95 per cent humidity, our noses have to do the work of translating. And in different environments this can be a different job. In desert areas the air is already heated, but not moist enough. Noses in these areas tend to be large and narrow, so that air travelling up

the nose has plenty of time to absorb moisture from the mucous membranes. In northern Europe the air is more humid, but colder. Long narrow noses restrict air flow and give the air time to warm up before reaching the lungs. In more humid environments the short wide noses common to many African and Asian people are more efficient – the air arrives at the lungs more quickly.

The process of evolution that led to different nose shapes was attributed by nineteenth-century travellers to fashion. European people speculated that oriental mothers lay their children on their front to compress their noses, all for the sake of beauty. And Tahitians, observing the longer noses of English children, lamented the supposed habit of English mothers of pulling their noses to make them long!

LIPS

The lips are the most visible feature of the mouth. They have a darker, more pink colour than our skin because they are so thin that the blood beneath the surface is visible. Our lips are extremely sensitive, and can detect a single hair in our food, as well as enable us to enjoy a kiss!

TEETH

The enamel of our teeth is very hard, and so they are the longest lasting feature of our face. We and other mammals have a more complex variety of teeth than other animals. Our incisors work to cut pieces out of food. The canines, all that is left of our fangs, are deeply rooted in the upper jaw, and are for gripping and ripping tougher foods. The premolars and molars crush food into a pulp, which makes it easier to digest. The wisdom teeth, which are behind the molars, often never emerge. It is possible that they are being 'selected out' by the evolutionary process, for with our pre-prepared and cooked food we no longer need so many teeth.

Previous pages Contrasting skin colours and types.

Opposite The concentration of melanin in the skin usually corresponds to that in the iris.

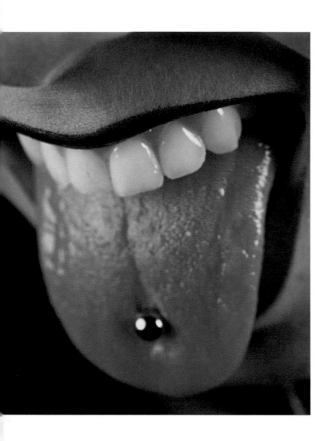

TONGUE

Our tongue is largely hidden from our facial display, but is an extremely important feature. It is formed of a bundle of muscles, which is why it is capable of such remarkable changes in shape, width and length. The tongue forms many of the shapes we need for spoken language, registers different tastes on the taste-buds, and is used when we swallow. We tend to swallow about once a minute – except when we are eating, when we swallow about nine times a minute.

CHIN

Our chin is a prominent – sometimes very prominent – feature, and yet is a puzzle for scientists. It has no clearly defined function. Earlier in our evolution we had a muzzle, which thrust our teeth outwards, meaning they could be used as weapons in hunting. But now, our teeth are too far back in the skull to be much use as weapons. The need for such a strong biting action gradually became redundant once we had discovered fire – hundreds of thousands of years ago – and were able to cook our food and tenderize it. As we saw earlier, the chin simply seems to be the remnant of our once larger jaw.

So our facial features have evolved in response to the particular environmental challenges of different areas of the world – as well as 'sexual selection', in which aesthetic preferences for certain faces varied from tribe to tribe, and was expressed in their choice of sexual partners.

However, these features differ from one person to another, as well as ethnically. Our diverse appearance around the world is made all the richer by the increasing degree of racial mixing. Today each of the myriad human faces in the world is a mass of subtle genetic mutations and developments. We each have the basic equipment – two eyes, a nose, a mouth, two ears – and yet we all manage somehow to look individually different, which is extraordinary.

So let us look next at our individual identity, and see how our faces help us to understand who we really are.

Opposite Two celebrities that have been known for their prominent chins – Bruce Forsyth and Jimmy Hill.

Overleaf Surma woman in Ethiopia. The huge lip plate is made of clay.

identity

CHAPTER 2

THERE ARE SIX BILLION HUMAN FACES ON THE EARTH AND, REMARKABLY, EVERY ONE OF THEM IS UNIQUE. The face we see in the mirror every morning is ours alone. That is why our face is nature's name-tag. We know who people are by their appearance, and that is how they recognize us.

Where do our looks come from?

Next time you're near a mirror, have a good look. Not the usual quick glance just to make sure you are still there. Or to check your make-up. Or to see if the latest disastrous new haircut is growing out yet. Instead, consider your features. You are the only person in the world who looks exactly as you do. Where did those looks come from?

The genes your parents passed on to you were inherited from their parents, and so on down the seemingly infinite line of generations. So your unique looks are the result of a distillation of the features of very, very many people.

We all know families where the facial resemblance between the members is strong. But science has yet to unlock the secrets of the genes that construct the face. Facial features are the product of multiple genes, and trying to find the combinations that give you a button nose or an aquiline one is proving to be a difficult code to crack. It is that complex interaction that creates such an immense number of unique faces. So far no one gene has been found to be directly responsible for a single facial feature.

Genetic inheritance is passed on from generation to generation in the form of twenty-three pairs of chromosomes, one member of each pair coming from the mother and the other from the father. Genes, strung out along the chromosomes, therefore also come in pairs. Within each pair, one of the genes may be either dominant or recessive with respect to its partner, so that only one of the genes may be expressed in the offspring. This means that some of your features may have lain dormant, and not appeared on anyone's face until now (yes, we know, it's really bad luck…). The particular complex combination of genes that conjured up your face is interacting now for the first time.

Above Every baby's face is the result of a combination of multiple genes, inherited from both mother and father.

Opposite As we grow older, so our face changes, partly due to genetics, and partly due to the influence of environment.

Above Getting older does not
necessarily mean losing your looks.

Opposite Male and female faces
differ significantly – female skin tends
to be lighter, and the male has a
larger and wider nose.

Why are male and female faces different?

There is an aspect of our face that we may wonder about in childhood, but by adulthood we seem to take for granted. It is: why do I look like a man, not a woman? Or vice versa.

Babies might look pretty similar to everyone except the doting parents, but there are already differences between the faces of male and female infants. At birth the baby boy's head is bigger than his sister's. As they progress into adulthood the female skull grows to, on average, only two-thirds the size of the male's. There are other structural differences too. The male head has bigger ridges of bone on the jaw and brow. He also has a larger, longer and wider nose. This is partly to do with being a bigger creature – he needs to suck more oxygen into his lungs in order to carry the extra bodyweight.

We become most aware of these gender differences during adolescence. Young women's lips appear fuller than those of her male counterparts. Puberty causes male skin to coarsen. Testosterone makes pores expand and facial hair sprout – the easiest way to tell men and women apart. Men's eyebrows are thicker.

Women have more fat on their cheeks, giving a plumper, softer outline. High cheekbones are a myth – it is nothing to do with bones, but rather the way the fat is distributed under the skin. Greta Garbo famously had teeth removed so that her cheeks would recede, making her 'cheekbones' look higher.

Left to right Women of all races usually have lighter skin than the men; actress Greta Garbo; Margaret Rutherford – as a woman ages, hormone changes make her look more masculine.

In contrast, women's faces do not have as much bone mass, but their eyes are more prominent. The eyeball is practically the same size in men and women – but because women's eyebrows are higher up, their foreheads more upright and their brows slight, the eyes look larger. And that's even without the benefit of eyeliner!

Why might we need to tell men's and women's faces apart? In evolutionary terms the answer, as usual, is sex. We need to know whether or not that person is a possible mate. That is also why male and female faces only really start to look different from adolescence onwards – before the onset of sexual maturity it really does not matter to Mother Nature whether we can tell Jane and John apart. This sexual function can be seen in differences in skin tone. Women usually have lighter skin than men of the same race. Female skin lightens at puberty when the levels of melanin and haemoglobin go down and oestrogen rises. The lighter skin colour of females may also be an indicator of fertility. A pregnant woman's skin darkens, indicating her condition.

It is the hormone oestrogen in females that inhibits the adoption of male facial characteristics. As women get older, and become less fertile, the ratio of the male and female hormones alters and their faces begin to look less feminized. The lips shrink in size and fullness and the eyes start to look smaller.

A man who spends five minutes a day shaving, starting at the age of fifteen, will by the age of sixty have spent an estimated 1368 hours of his life on this chore. That's nearly sixty days and nights of nonstop shaving.

Above Chimps are one of the few creatures that can recognize their own reflection.

Opposite A child of around 18 months of age is able to recognize him- or herself in a mirror.

Recognizing ourselves

Seeing our own reflection is something we take for granted, and is a fundamental aspect of our self-identity. But it is an unusual ability in nature. We have to go a long way up the evolutionary ladder before we find a creature that recognizes its own face in the mirror. Cats do not – they peep around the corner of the mirror, and slink away. And dogs don't either – they completely fail to see their reflection, or just lose interest in the experiment. Even most monkeys do not recognize the image of themselves – they either treat it as friend, and get playful, or foe, and become threatening, until they tire of it.

Chimps, however, realize that the reflection is of themselves. At the State University of New York, psychologist Gordon Gallup anaesthetized chimps and, while they were asleep, painted odourless red lines above their eyebrows. When they awoke, he set a mirror before them. They all tried to touch their foreheads, patted the red strip in puzzlement, and many then smelled their fingertips. This demonstrated that the chimps recognized their images, since otherwise they would not have checked their own bodies. He later found that orang-utans, gibbons and gorillas show the same reaction to a red stripe on their head.

For humans, it takes a little while to get the hang of it. As one-year-olds looking in the mirror we see not ourselves but another person. But after that age, our understanding of our environment begins to become more sophisticated. Usually by 18 months, when we see our reflection in a mirror, we will know immediately who it is. We no longer assume that the image is somebody else who just happens to look like us. We are able to monitor movements in the mirror and realize that it is we who are making them.

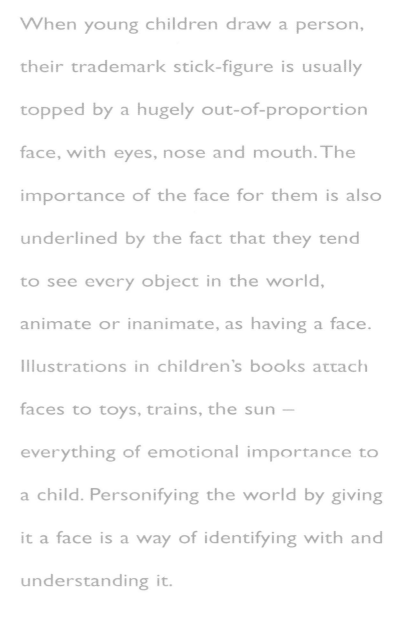

When young children draw a person, their trademark stick-figure is usually topped by a hugely out-of-proportion face, with eyes, nose and mouth. The importance of the face for them is also underlined by the fact that they tend to see every object in the world, animate or inanimate, as having a face. Illustrations in children's books attach faces to toys, trains, the sun — everything of emotional importance to a child. Personifying the world by giving it a face is a way of identifying with and understanding it.

Once we have recognized our reflection, there is no stopping us. We gaze at ourselves in a mirror thousands of times in our lives. Our regular and familiar appearance in the mirror does much to reassure ourselves about our identity. Anyone who effects a significant change in their appearance – a distinctive new haircut, for instance – will know what a shock it can be to forget about it, and then encounter oneself suddenly reflected in a mirror or shop window. A physical transformation of our face powerfully affects the way we experience ourselves, and the ways in which others respond to us.

While the ability to see faces is inbuilt at birth, the ways in which those images present themselves can take some getting used to. When anthropologist Edmund Carpenter gave Polaroid photos of themselves to tribespeople in New Guinea, they were puzzled by the little bits of card he handed out as gifts. They had never seen two-dimensional images of themselves like this, and Carpenter had to explain, finger pointing to a feature such as the nose in a photo, and then the nose of a person.

Suddenly they could recognize themselves. They were shocked – some ducked their heads and turned away, came back for another look, ducked away again. Eventually they could hold their gaze on the picture, mesmerized, trembling with tension. Some slipped away, photo clutched to their chest, to find a private place for closer inspection. Others wore the Polaroids stuck to their foreheads. These people were confronting their visual image for the first time, and were stunned not only by its novelty, but also by the potency of such an icon – a representation of themselves that perhaps carried the power of their soul or essence, their deepest identity.

When we look at a face in a photograph, our own face in reflection, or others' faces in front of us, how do we go about putting together the splodges of shadow and light? We can learn something of this process from someone who cannot recognize faces.

Jim Cooke lives in New York State. He has normal eyesight, but he cannot recognize his own face in the mirror. He has to shave by feeling his way around his features. In 1995, then aged forty-eight, Jim went in for a brain operation. When he came round, he realized something was wrong. It took him several days to figure out that he could not see faces properly any more. Jim was now suffering from prosopagnosia. This term is derived from

Previous pages Is this man looking at someone he knows, or at a stranger?

Opposite We are able to pick out a face we recognize amid a crowd of strangers.

Opposite Our iris pattern is unique, a method of recognition even more accurate than fingerprinting.

Below Middle-aged male twins.

Overleaf Young female twins.

prosopon ('face') and agnosia ('lack of knowledge'), and refers to a disorder in which people can see most things normally, but when they look at a face they see only a canvas of features that do not form a meaningful image. They cannot recognize the face as someone familiar.

You can be born with this rare illness, but more commonly it occurs as the result of adult brain damage. While the brain as a whole is involved with most perceptual functions, there is a tiny, specialized section of the brain that is intimately involved with the recognizing of faces. It is called the fusiform. In a scanner it lights up with electrical activity whenever a person looks at a face. If the fusiform gets damaged we cannot register the face as a face – and Jim's fusiform area was damaged during the operation.

Jim explains that when he sees faces: 'It's almost as if everyone's wearing stocking-masks.' It is disturbing for Jim if people recognize and approach him in the street, since he has no way of placing them or guessing who they are. Most of us forget the odd name or face, but Jim can't see any faces at all. Most distressing for Jim is the fact that he cannot respond to the faces of his own children. When he goes to meet his twenty-year-old son Tommy, or his eighteen-year-old daughter Cindy, he cannot recognize them. Tom and Cindy have learnt to cope with their father's illness. They make sure that they say 'Hi, Dad' or identify themselves every time they approach him.

Prosopagnosics evolve complex strategies to deal with their illness. They become expert in differentiating voices and clothing, so as not to give away their problems. But for anyone, this is an enormously debilitating illness.

Jim Cooke says that: 'When I look in a mirror, I'm not there. I'll see items on the wall behind me, but a blank in the middle...' As an adult, to be unable to see our own face while being able to make out everything around, is like being robbed of a key to our identity. Jim Cooke's experience underlines the importance to us of face recognition.

Systems such as identikit, and the more modern E-fit, ask us to build an image of a face systematically, feature by feature, by selecting the eyes, then the nose, then the lips, in strips. This process helps us to remember faces we have glimpsed, but it is not how we recognize a face when we see it. We identify faces by integrating two principal sources of information: patterns of light and shade on the face, and the proportions or relationship between the

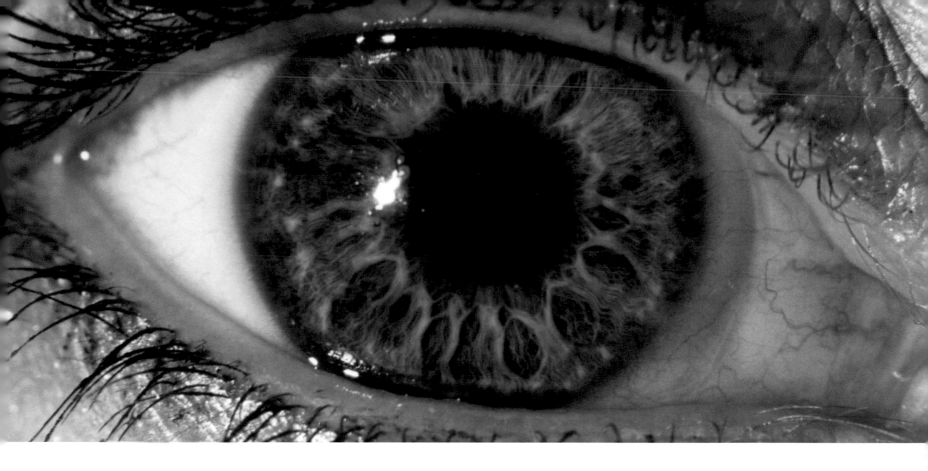

Telling people apart by their facial appearance is usually easy. It is much more challenging when we are faced with identical twins. Identical twins have identical DNA. They are clones in every way, yet they do not look absolutely identical. In the finest detail their faces, like their fingerprints, are distinct. As they grow out of infancy their experiences begin to exaggerate the subtleties of their appearance and parents and close friends learn to tell them apart. Technology allows us to look closer. The IriScan takes a picture of the iris and plots 266 points on it. This method of recognition is even more accurate than finger-printing.

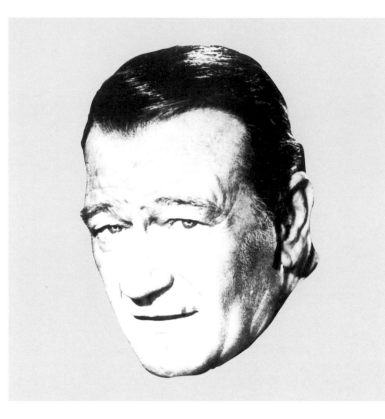

Above It is difficult to recognize even famous faces like John Wayne without the detail of light and shade.

different features. Our brains recognize faces as meshes of shadows and shapes, and as a whole rather than as separate bits.

Knowing the proportions alone is not sufficient for face recognition. Psychologist Graham Davies investigated this by tracing around the features of photographic images of famous faces. He then showed these detailed line drawings to observers. The observers were able to identify only 47 per cent of the people from the line drawings, compared with 90 per cent when they were shown the original photographs. If our memory for the faces of people was based only on features, such as the length of nose or width of eyebrows, then we would be able to recognize the line drawings as effectively as the photos. To accurately recognize the famous faces, the observers needed the patterns of light and shade offered by the photographs.

Light and shade gives us a necessary three-dimensional image of the curves, bumps and detailed contours of a person's face. However, line drawings of a different kind deal with another important aspect of face recognition: distinctiveness. How a face departs from the average or prototype helps us to register and recognize it. Caricaturists have for a long time been drawing people this way. Exaggerating, sometimes extremely, the most characteristic features of a well-known politician, for example, seems to render that person almost instantly recognizable in the drawing. University of Southampton psychologist Sarah Stevenage reviewed scientific research in which the caricaturing is done by a computer, simply emphasizing the features along the dimensions on which the face is naturally distinctive. The research shows that we seem to be able to read more from these exaggerated drawings than the tracings that reproduce the features accurately. The exaggerated lines trigger recognition better than literal sketches.

Our face is such a familiar sight. Sketching the face allows us to see ourselves, and others, in a new light, if done in the right way. Or rather, the wrong way.

Take a mirror and place it so that you can see yourself easily. Set a sheet of drawing paper in front of you. Place your pencil point on the paper, ready to draw. Look up at your face, and begin to draw it. But there is an important rule to be followed in this exercise. When drawing your face, at no time must you look at the sheet of paper. Do not take your eyes from your face in the mirror. Then carefully try to draw the outline and features of your face without lifting your pencil from the paper. Do not look down until you feel you have finished.

You might wonder, if there is not constant checking between the object being drawn and the lines on the page, how you can hope to have an accurate representation of the face? That is the very point!

The task is to sketch an outline like a caricaturist, which captures the essence of a face. This necessitates fresh perception — recording details free from the habitual assumptions we make in forming brain images. For the sketches to be any good, we need to allow them to be terrible in conventional terms. The images will not look at all like the literal, 'still life' drawings we did at school. They look like a Picasso. Strong, vibrant, impressionistic; inaccurate in some details but sometimes stunning in the way they capture the essence of a face.

Recognizing strangers

Below Some of the many illegal passport photos of drug trafficker Howard Marks. Would you recognize him?

'Can you confirm that this was the man you saw stealing the gold watch from the jeweller's window?' Sometimes our ability to 'recognize' a face is put to the test in making important decisions – like when we are a witness to a crime.

While we are good at recognizing people we know, even in difficult circumstances, it is a rather different story when we try to identify people we have seen just once, briefly. In the laboratories we can be good at playing snap with faces. At Stirling University, Vicki Bruce has researched the processes involved in face recognition. She describes studies in which subjects are shown hundreds of mug-shots, and then re-shown them a couple of days later; people were as much as 96 per cent accurate in remembering which ones they had seen before. But if the pose or lighting is altered in the photograph from first viewing to second, even though the images are of the same people, recognition drops to about 60 per cent – scarcely better than chance. Recognition seems to be a sensitive, fairly primitive skill, based around patterns of light and dark.

Opposite We rarely take notice of strangers when we pass them.

It is not surprising then, that in real life, we very often get it wrong. Generally we get little practice at scrutinizing and processing people's faces. There are some professions – the police, for instance – where people are trained, and may become better than average in attending to and recognizing faces they have seen only once before. But in general, we rarely look closely at strangers as we walk down the street, perhaps because in many cultures this is considered rude.

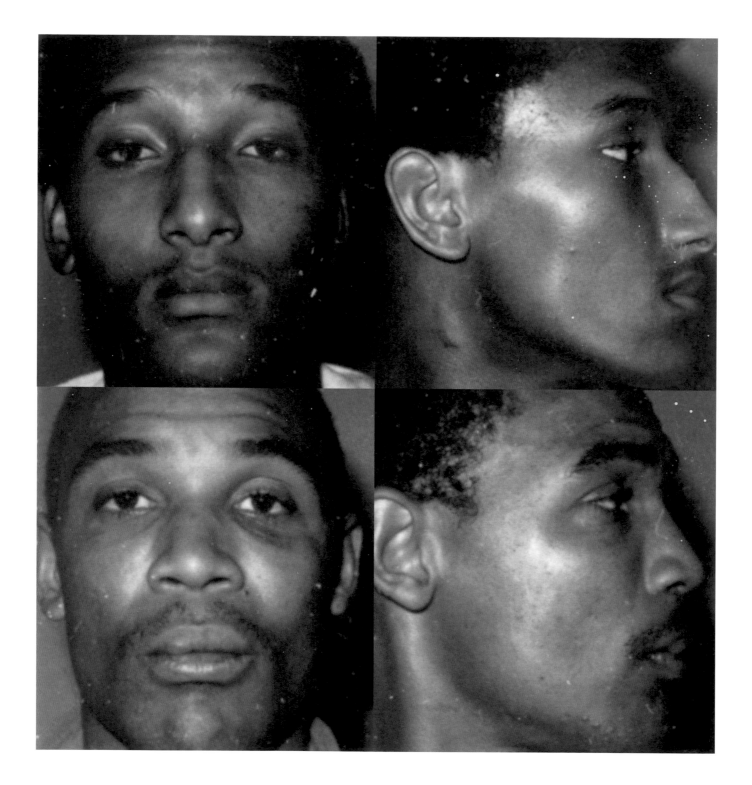

And so people turn out to be poor at recognizing faces they have glimpsed only once. And things get worse if we are confronted with faces of a different ethnic type, unless we are familiar with such faces through extended contact.

One such disturbing story concerns a woman named Jennifer Thompson. In 1984, as a young, white student in North Carolina, she was raped by a black man in her home one evening. She was determined to remember his face. Jennifer constructed an identikit picture of him at the local police station, and she identified the suspect in the police line-up. She was certain she had recognized her attacker – a man called Ronald Cotton – and the confidence of her identification was a key element in his conviction. Cotton always pleaded his innocence.

After several years in jail, Cotton encountered a man who admitted to being Jennifer's real assailant. His name was Bobby Poole. At a retrial, faced in the courtroom by Poole and Cotton, Thompson was absolutely certain that Cotton was the rapist. Cotton was sent back to prison. It was only at a subsequent appeal, when DNA evidence was introduced which proved that Poole – now dead – was the real rapist, that Cotton was finally acquitted and released in 1995.

For many years, Jennifer saw the wrong face in her nightmares. Even when faced with apparently incontrovertible evidence that our memory is mistaken, we can be so sure that we would swear that our memory is serving us right.

So if humans are bad at it, perhaps machines will be better? It turns out that machines are quicker and more accurate at registering and remembering faces than we are. They are able to match up photos and images by arithmetically plotting points on the faces. Various systems are already in use round the world. In Newham, East London, a system linked to CCTV can pick out people known to police on the streets by matching faces to mug-shots. A photo of the person in police records is scanned into the machine, and if the camera 'sees' the person on the street it can identify them.

But in general, the difficulty of identifying faces after only glimpsing them once has led to some appalling mistakes in court. Much of the recent research interest in the processes of human face recognition was stimulated by some high-profile cases of mistaken identity.

Opposite Rape victim Jennifer Thompson was convinced that Ronald Cotton (top) was the perpetrator. However, Bobby Poole (below) later admitted to the crime.

Recognizing family and friends

Not surprisingly, we are much better at recognizing people we already know – from close members of our family, to friends and acquaintances, even to people we have never met but who have familiar features, such as TV and film stars we have watched repeatedly. It turns out that we are good at recognizing all of these people even under difficult circumstances.

This ability starts very young. We learn to recognize our mother's face, and other members of our family, before we become familiar with our own. As infants, our vision is very poor – even at a distance of 30 centimetres we are very short-sighted, and can make out only smudges and shapes. But we learn to connect the pattern of shapes to a single individual – usually our mother – within days of birth. And this focus on familiar faces continues throughout our lives. Even fifty years after graduating from high school, people can accurately identify faces taken from their own school yearbook when compared with photos from others of the same era.

A person in a poor quality photograph or fuzzy piece of video footage is clearly identifiable if it is someone we are used to seeing on a daily basis. It seems that we can each describe in detail, or recognize easily, anywhere between about fifty up to several hundred faces. Recognizing people we see repeatedly entails making allowances for variations in their personal appearances. Age especially alters our looks. Even after not seeing someone we know well for weeks, months, even years, we are usually able to recognize them. But what if the absence is six years? And the person is your child, who has aged in the meantime from three to nine years old? Could you pick your child out of a crowd of children? This is the dilemma that faced Mirna Martines.

In October 1993 in Texas, Mirna Martines returned home to find that her three-year-old son, Johnny, had disappeared. She was frantic, but suspected at once that his father, worried that he would not be awarded custody of Johnny in an on-going custody battle, had decided to take matters into his own hands and kidnap his son. There was nothing Mirna could do. The police turned up no evidence that could locate her son. The years passed.

Mirna tried to get on with her life – she remarried and gave birth to another little boy – but was living with the pain of not knowing if her missing son was alive or dead. Eventually she went to the National Center for Missing Children in Washington clutching a photo of Johnny. They used a cutting-edge computer technique to track him down. They scanned the photo of Johnny at age three into the computer and then artificially 'grew' his face, morphing it on the computer to reflect the changes in size and shape of his features characteristic of a boy of nine, as he would now be. Then they scanned in childhood photos of Mirna, and other genetically related family members at a similar age, to provide some family resemblance background. They further manipulated the image until they were satisfied that they had something that might look like the real Johnny. It was the first time Mirna had 'seen' her son in six years. But there was still a long way to go.

By now, the police had tracked down some relatives of Johnny's father in Texas. They suspected that if Johnny were still alive, this is where he might now be living. The National Center sent posters of the computer-aged Johnny out to supermarkets in the Dallas area. Within weeks Johnny's photo was everywhere. Then one day there was a breakthrough. A kid and his mum walked into their local shop and the boy turned to his mum and said, 'Oh look, that's Hubert Dias.' Could this be the pseudonym that Johnny's father had given him? The police interviewed the boy, who had played with Johnny/Hubert, and narrowed down their search. They ended up at the McLively Elementary School where they showed the computer-aged photo to the teachers. Again the same name came up. The police staked out the school and, sure enough, Hubert arrived and on questioning turned out to be the long-lost Johnny! He was reunited with his mother in August 1999.

Opposite Familiar faces. Close family members would instantly recognize Claire, the grandmother, despite changes such as the 30s fashion for plucked eyebrows. But would she have recognized her granddaughter, Erica, if they had lost contact after her birth and not met again until Erica was a teenager?

Below UK toddler Ben Needham (left), who went missing aged 21 months during a holiday in the Greek islands in 1991, and (right) the computer-'morphed' image released by police of Ben as he might look nine years later, in the hope that he can still be recognized and traced.

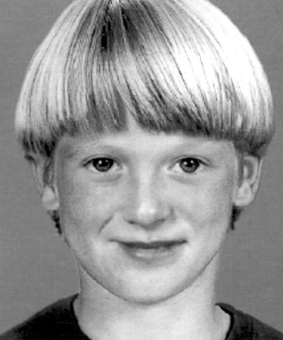

All of us are changing, all the time. Our features age, and reflect our physical and emotional states of health. We can see those gradual changes in the mirror. But we still feel like ourselves. So our face is our identity tag, for ourselves and others, but a dynamic one. It has recognizable features, but it moves with the times and reflects our journey through life. It is said that 'after

Right and opposite The faces they deserve? The ageing process, reflecting two people's journey through life from birth to middle-age.

forty, we have the face we deserve'. Our features reflect the way we feel and the way we live, because over the years our faces come to show our characteristic expressions and habitual visage. In the next chapter, then, we shall explore the nature of this expressiveness and consider what it tells us about our ever-changing selves.

expressions

A GENUINE SMILE GIVES US A WARM GLOW OF PLEASURE. A quick raise of the eyebrows grabs our attention – it is our most common expression of greeting. We might not want to tell the person we're flirting with that we are attracted to them, but our eyes will give them the message. And a blush says a lot about how we are feeling. We are all sophisticated and sensitive readers of other people's expressions.

Equally, a frightened face mobilizes our own alarms immediately. Seeing eyes wide with fear, even when expressed by an actor on film, makes our heart rate shoot up, and we begin to experience the sensations that the scared person feels. The empathy is real, biological and strong. Facial expressions seem to be one of nature's inbuilt ways of communicating with others on a deep and immediate level.

Of course, most animals use body language to complement their cries, song and sense of smell. The natural world is a babble of vocal and physical signals and complex codes. However, not only have humans developed their cries and song into a uniquely elaborate system of speech; we have also, more

than any other animal, exerted our control over facial muscles and thus expressions to a highly sophisticated art.

We might think that facial expressions are redundant now that humans have language, an infinitely superior tool for expressing ideas and concepts. But in evolutionary terms, language is a relatively new addition and has its limitations. Our faces can express things that are difficult to put into words. Expressions can communicate emotions faster, more subtly and more effectively than words, which is why facial expressions remain crucial for humans as social animals.

And it is also why facial expressions are the analytical language behind the 'Love Lab'. Since 1990 psychologist John Gottman has studied almost 700 married couples in the lab at his research centre in Seattle. The partners come for a 'diagnosis' of their emotional relationships. They are videoed as they sit in a room discussing with each other distressing issues in their relationship, or reminiscing about how they met. Afterwards, Gottman analyzes their

The human face is the most expressive in the world.

Left The Princess of Wales – shy of the cameras as the young Lady Diana; as Princess of Wales, still uncomfortable with public attention; shortly before her death, openly discussing her private life on BBC1's *Panorama* programme.

Above and opposite Relationships in trouble.

Expressions are more personal than speech, because they communicate our feelings.

non-verbal behaviour on the video in great detail. Focusing on their facial expressions, he identifies and counts fake smiles, and compares them with the genuine smiles by watching for the telltale creasing around the eyes that show the smile is not 'manufactured'. (There's more about real/fake smiles later in the chapter.)

He also looks for people's facial expressions of contempt for their partner, sometimes so fleeting that they can be seen only on stop-motion video. The 'dimpler muscle', or buccinator, which pulls the lip corners to the side and creates a sneering dimple in the cheek, is often combined with a slight upward eye roll. Gottman has found it to be a potent signal of future marital disintegration.

John Gottman sees people who are in successful relationships in his lab, too. Over the years he has kept in contact with all his participating couples, and monitored the course of their marriages. And he is now able, after analyzing only three minutes of interaction on video tape, to predict at a startling 75 per cent rate of accuracy whether a couple will get divorced within six years. Give him fifteen minutes of videotape, and his accuracy rating climbs to 85 per cent.

John Gottman achieves these rates of prediction not by listening to what couples are saying, but by analyzing their body language and facial expressions. Non-verbal expression is more revealing of emotions than verbal interaction. In speech we are able to cover up, to conceal, to lie – even to ourselves – about how we really feel. We try to cover up our emotions in our expressions, too. But some expressions are made unconsciously, and even the many that we can control voluntarily are difficult to manipulate and are amenable to analysis and diagnosis.

We use expressions all the time. Just walking down the street we are inundated with non-verbal cues from people's faces. We know from experience that when we are with people we are interpreting their facial

expressions constantly. Of course, successful communication between people occurs not when the expression is sent out, but when it is correctly understood. We try to gauge what other people are 'saying' by their expressions, and we have to do it without the help of stop-motion video analysis. We also have to take into account that some of their expressions are meant to conceal what they are thinking and feeling. So interpreting the meaning of each other's expressions is a complex process, and not surprisingly we make lots of mistakes.

In order to explore this finely balanced world of revealing and hiding our emotions, we first need to consider the underlying mechanics of making expressions. The physical characteristics of the face reveal something of the secret of the heartwarming smile, and the sneering dimple of contempt.

How expressions work

Some people seem to be more expressive than others. Their faces are animated. Their emotional lives seem to bubble to the surface readily. In contrast, others keep a tight grip on how much they 'give away' of their feelings. Research suggests that people who are positively expressive may also be happier. This sounds like a truism, until we take on board the time scale involved in this study.

As part of a long-term health and psychology research study, Dacher Keltner interviewed a number of women thirty years after they had stood together posing for their high school yearbook photos. He found that those who appeared in the photos to have natural, sincere smiles on their faces reported that they were in happier relationships, three decades later, than the ones who had looked unsmiling in the yearbook. Perhaps this reflected an underlying propensity to happiness, an optimistic attitude to life that endured over a lifetime. But it may also be that there are advantages to having an expressive face, and that this ability to communicate one's feelings contributes to successful relationships.

In assessing ability to identify the expressions on the faces in a series of photographs, young children who were best at spotting the correct meanings — like who was upset, and who was happy — were also found to be the most popular within their peer group. We all like people who are sensitive to how we are feeling.

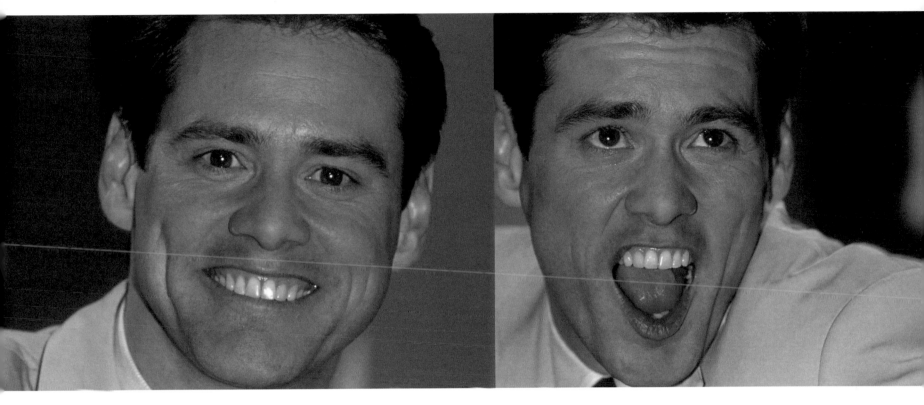

Sadly, the option of smiling is not open to everyone. In Fredericksburg, Virginia, when five-year-old Lauren Deveney had her photograph taken, she could not produce the 'yearbook smile'. Her face always looked impassive. She has a distressing condition called Mobius syndrome, which leaves people unable to smile, or even blink. She is one of a small number of people who are born without the cranial nerves that operate facial expressions, leaving them incapable of making any expressions whatsoever.

The resultant blank expression can lead others to make incorrect assumptions about the emotional states of people with Mobius syndrome. For example, teachers may feel that they are not paying attention, not showing any interest in the class, or do not care about their work. Lauren's parents, Dan and Sharon, were also concerned that her condition might leave her cut off from emotional communication with others and that, as she became older, she would be excluded from a rich area of sociability that the rest of us take for granted.

Above Actor Jim Carrey has forged a successful film career with the help of his very expressive face.

Mobius is an incurable disease, as doctors are still unable to reactivate dead nerves. But pioneering surgery offers children like Lauren a chance. And things have changed for her, for early in the year 2000 she travelled to Toronto to undergo surgery to enable her to smile. The procedure involves taking a portion of the gracilis muscle 3–4 centimetres long from the inner thigh, and implanting it into the cheek. It is then attached to the biting nerve, so that the patient can learn how to smile by biting down on the back of her teeth. One side of the face is operated on first; about eight weeks after the surgery, initial movement in the face begins. All being well, the second operation may take place three months later.

Lauren underwent both stages of the surgery successfully – in time for her to start school, complete with a smile.

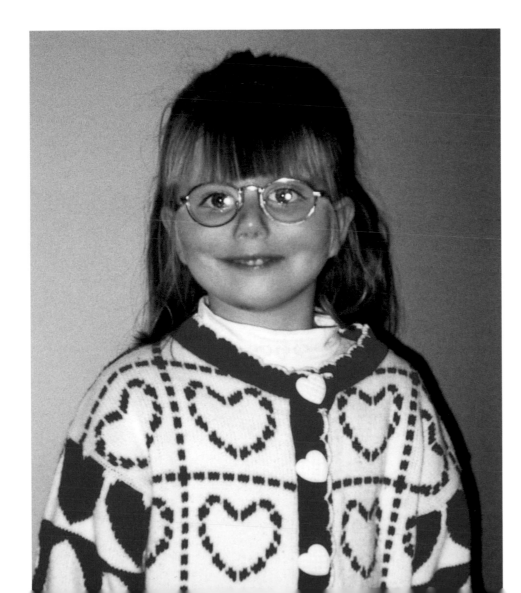

Opposite and left Lauren Deveney has Mobius syndrome and underwent an operation to enable her to smile.

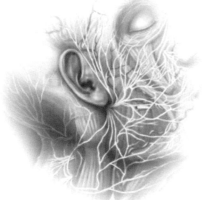

Above The facial muscles form a complex mesh beneath the skin, and move against each other to form our expressions.

This story reminds us that behind the analyses of meaning, facial expressions are essentially a physical action. They depend on very complex co-ordination of nerves and muscles in the face. And the human face is especially designed to make this possible. Beneath the skin, our face is covered by closely intermeshed muscles, all lying across one another. This is quite different to how muscles are arranged in other parts of our body. There the muscles begin and end connected to bone, and they contract in order to move our joints. In the face, many of the muscles begin and end within each other. They form a matrix of moving tissue just under the skin. When the brain transmits its messages to the facial muscles via two main nerves – one controlling the facial expressions, the other controlling blinking – these muscles contract or relax and move the face to form our expressions.

The muscles on our face enable us to make up to 7000 distinct expressions. But just like our spoken vocabulary, in normal day-to-day life we only use a small proportion of the total – a few hundred expressions.

We send signals from our brain to activate the expression of emotions. Many parts of the brain are involved. But as we have already noted, in evolutionary terms expressions are a more ancient way to communicate than spoken language. So while speech is processed in a relatively recently developed segment of the brain, a basic emotion like fear is picked up and processed in non-conscious centres of the brain, such as the tiny amygdala.

The amygdala function as an alarm signal. Magnetic resonance imaging scanners, which can map a person's brain activity on a computer screen, show that the amygdala lights up with brain activity if confronted with, for example, an image of a fearful or angry face.

Opposite A face of fear: eyes wide, mouth gaping open, ready to emit a scream of terror.

Sometimes this basic response to a potentially alarming situation fails to work. One recent theory is that people diagnosed as psychopaths – who engage in socially destructive behaviour with no apparent remorse – have a malfunctional amygdala. Scans show their amygdala making little response to the sight of another's distress, or to signals of threat. As a result they have chronic problems reading the emotions of other people, and modifying their behaviour accordingly.

We are all susceptible to losing this safety loop from time to time. Scientists have found that alcohol similarly reduces our response to the emotional expressions of others, and the amygdala may be involved. When we are drunk we can fail to recognize 'alarm' expressions such as anger. This may help to account for the frequency of drunken brawls. If we cannot tell if someone is angry and so do not moderate our own behaviour accordingly, then we are more likely to get into aggressive confrontations with them.

Clearly our expressions are heavily reliant on biological mechanisms, some of which have been part of our neurological make-up since very early in human evolution. In fact, they are such ancient neurological structures that other animals must have similar brain functions. And yet humans are unique in the range of expressions we are able to create. How then did our expressive abilities evolve?

Our flexible face

Pity the poor crocodile. It can make only four facial expressions – variations on eyes open and closed, mouth open and closed. Because reptiles are cold-blooded animals, they are covered in the thick, rigid hide that keeps their inner body temperature stable. It also makes for a pretty immobile face.

In just the same way that we can trace the evolution of our faces back to the beginnings of life on Earth, we can also trace our expressions back to our ancient ancestors. As we saw in Chapter 1, our own distant reptilian ancestors evolved into warm-blooded mammals, and our faces changed considerably. Because our skin became thinner and more mobile, we developed more possibilities for non-verbal communication. Instead of tearing our prey apart with huge reptilian jaws, we were able to bite and chew our food into more

Above Humans have developed a more delicate way of eating, as we no longer have to tear flesh with our teeth.

manageable pieces. The mouth became smaller, the face became more flexible and, to our eyes anyway, a more attractive package.

In our early mammal incarnation, in which we resembled the shrew-like Megazostrodon, we relied on an acute sense of smell to find food and to communicate with fellow creatures, and on touch to negotiate our way around small spaces using sensitive whiskers. The long mobile hairs of our whiskers were moved by finely controlled muscles at the side of the mouth and nose. But once smell and touch were no longer our dominant senses, then the large facial hairs were redundant. We lost our whiskers and the muscles controlling them were now available for increasing our capacity for a variety of facial movement.

Our tree-inhabiting ancestors, 30 million years ago, were more visually dependent. Over generations, the positioning of our eyes gradually migrated to the front of our faces, giving us three-dimensional vision and making it less likely that we would miss our swinging vines and plunge to the forest floor far below. And crucially, improved eyesight meant that we really started to notice the facial movements with which our mobile-featured fellow primates 'expressed' their emotions and intentions.

Eventually we lost much of our facial hair. Our 'naked' face today allows us to see the tiniest muscular movements. Whereas it is pretty difficult among all that fur to see if a chimp raises its eyebrows, in humans the two lines of facial hair above our eyes have become dancing messengers because they stand out so clearly against our skin.

So our faces have developed remarkable mobility. We are able to show an enormous range of expressions. Yet within this flexible facial world, we are also able to recognize each other's emotions through the expressions we make. How is this possible?

Women's facial muscles are smaller than men's, and their facial fat hides them better. Yet women are more facially expressive overall, because they respond more openly to emotion than men do.

Following pages Grotesque contortion of the face forms the basis of British 'girning' competitions.

The basic expressions

The eminent nineteenth-century biologist Charles Darwin noticed that his dog, and other animals, seemed to have some facial expressions similar to those of humans. He reasoned that if such expressions were shared by humans and animals, then they might be biologically based, rather than culturally learned. Furthermore, they ought to be pretty universal for all people.

So Darwin set out to enquire systematically into expressions in various cultures. In 1867 he sent a questionnaire to missionaries and travellers all over the world, asking questions about the facial expressions seen among the communities in which they lived. The sixteen questions were along the lines of: 'Is astonishment expressed by the eyes and mouth being opened wide and by the eyebrows being raised?'

He sent out hundreds of these questionnaires, but in those days – before airmail, let alone e-mail – he had received only sixteen responses before the publication date of his book on the subject. Undaunted, and largely on the basis of his own ad hoc research, he published *The Expression of the Emotions in Man and Animals* in 1872. He concluded that expressions represented the same recognizable emotions all over the world.

However, after the initial success of the book, it was forgotten about in scientific circles, perhaps because the questionnaire was considered to be insufficiently objective. It was not until the 1960s that a young researcher unearthed Darwin's work and decided to put it to the test. In 1966 Paul Ekman, now Professor of Psychology at the University of California, took a collection of photographs of various facial expressions to an isolated tribe in Papua New Guinea who were still relatively free of influence from western culture. Ekman expected to find that Darwin had been wrong, and that expressions were culturally learned, and therefore distinct to each area of the world. He did not anticipate discovering that expressions were a universal language. Yet that is what he found.

Everywhere he went, the smile was always evidence of happiness, the frown of anger, and wide-open eyes of fear.

Opposite Dogs share with humans basic expressions such as anger and fear.

Following pages John Cleese trying out his expressions…

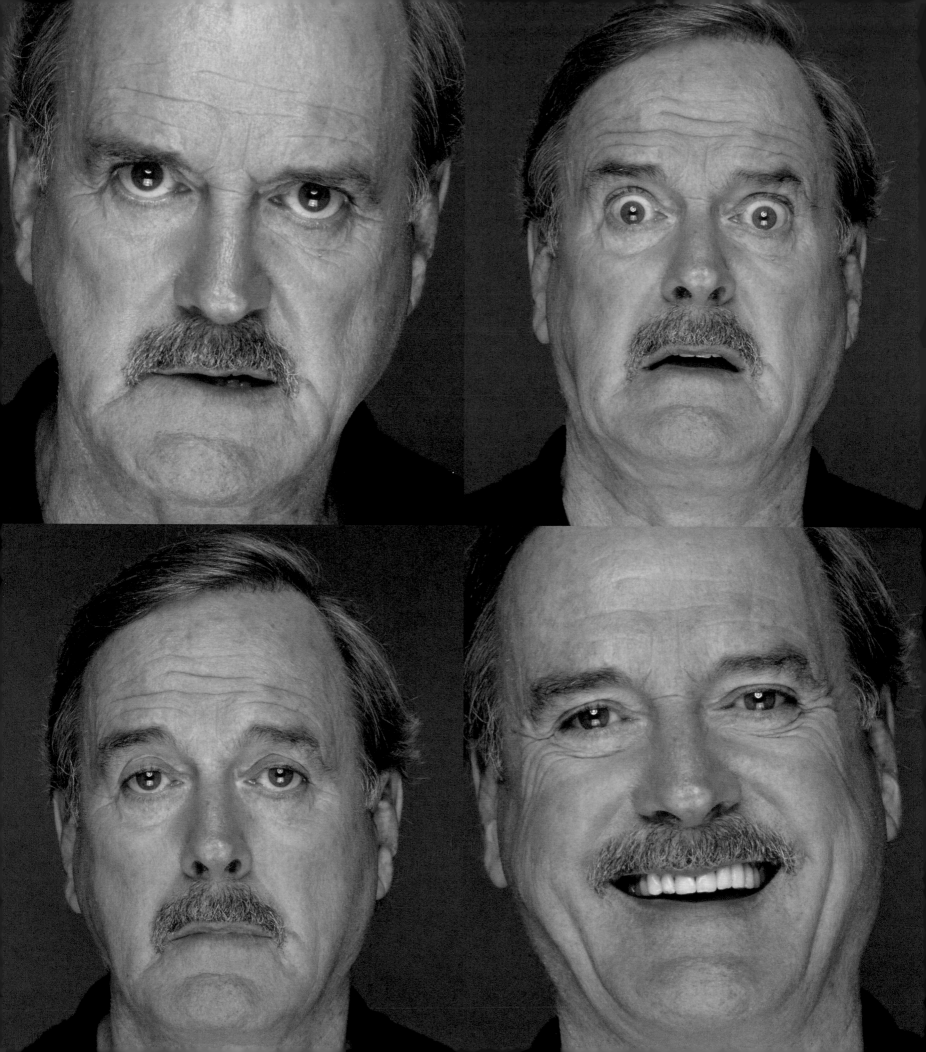

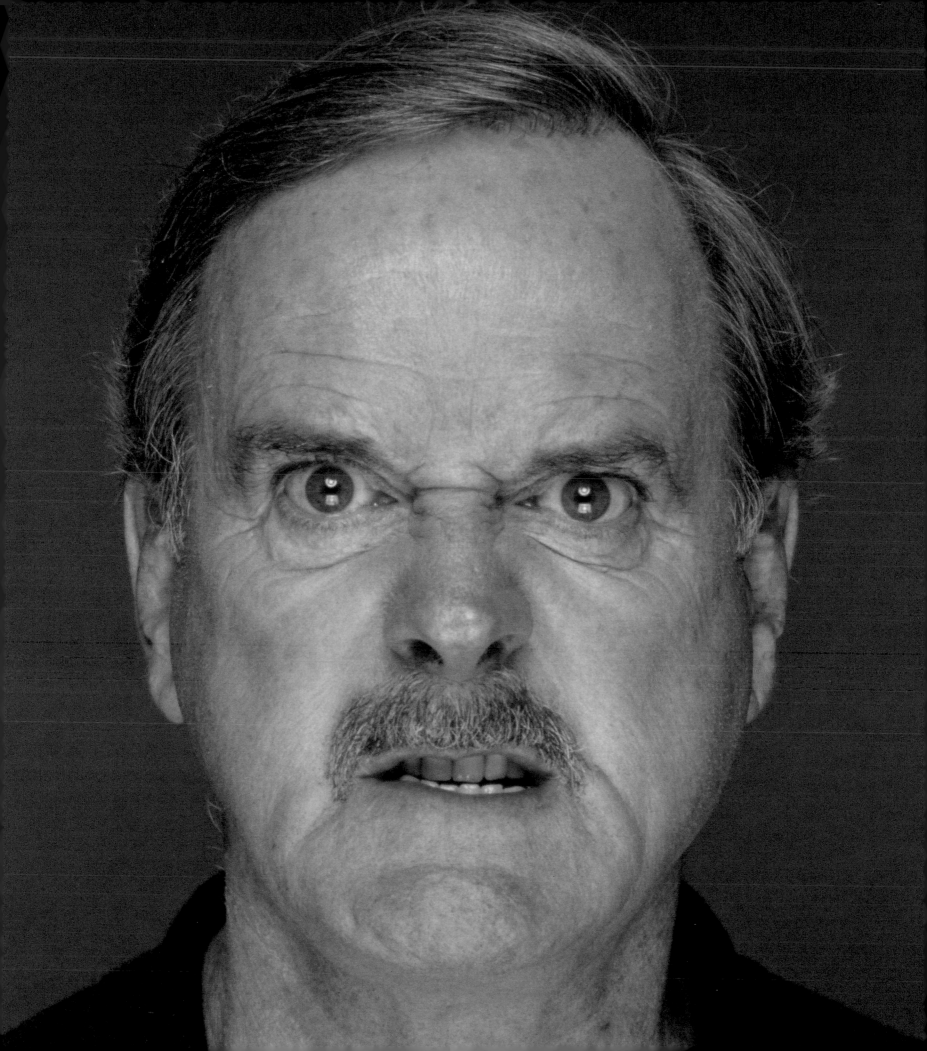

Above Babies revealing their feelings through basic, universal expressions: happiness (top) and disgust (below).

Opposite The 'eyebrow flash' greeting is used all over the world.

These findings set him swimming against the tide of psychological theory at the time, which assumed that all complex behaviours were learned, not instinctive, and therefore were culturally distinct. But over the years most psychologists have come to accept his results. Ekman suggests that there are six basic emotions that are connected with universally recognized facial expressions: anger, fear, happiness, sadness, disgust and surprise. Other researchers would add the expression of contempt to this list.

Although these basic, universal expressions reveal understandable emotions to anyone observing them, they are really more indications of an internal state than a deliberate attempt to communicate. If someone is surprised, angry, disgusted or happy, their face will register these emotions even if they are alone.

But we also know that there are many more expressions than these seven. Our everyday communication is full of subtle nuance, and rapid exchange of tiny non-verbal cues. In fact, contemporary researchers reckon we are capable of making up to about 7,000 discrete expressions. So how do we equate this with the finding that there are a small number of universally recognized expressions? Are there others that we learn?

Learning the variations

A baby is born with a fully functioning face that can make many of the movements associated with expressions. She can pucker her mouth around a nipple – essential for getting nourishment. She will frown and wrinkle her forehead in pain or displeasure and will wrinkle her nose in disgust if food is not to her liking. These expressions appear to be inbuilt in the human and are little more than reflexes. We know that these basic expressions are biologically based, as babies who are born blind develop a range of facial expressions. They still smile and frown without ever having seen anyone else doing so.

The basic expressions arise at fairly predictable stages. Disgust and distress, or sadness, show early – any time in the first three months. Babies start smiling as a reflex after about four weeks and the social smile, in which the baby responds genuinely to other faces rather than simply making a smile to exercise the face, at about six to nine weeks. Expressions of anger come on

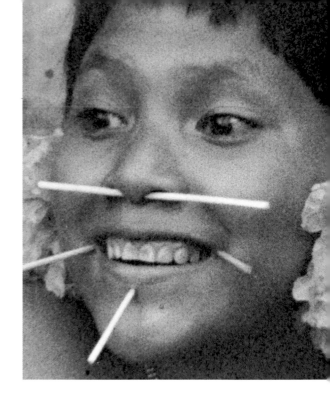

between three and seven months, and fear at five to nine months. All of these expressions show felt emotion, and are unlearnt. And yet adults, especially mothers, interacting with their infant children, seem to make special efforts to demonstrate to them a range of expressions – and infants seem to be primed to learn. Although the eyesight of newborns is poor – their focal point is just 30 centimetres from their nose and the picture they see is blurred, resembling what a very short-sighted person could see without glasses – they are nevertheless particularly responsive to expressions.

In the first days of life, the baby will start to mimic the lip and cheek movements of its parents. And a study of newborns showed that simple imitation of tongue protrusion and mouth opening of adults occurred after only forty-two minutes of life.

We respond enthusiastically to this extraordinary attraction babies have to our faces and alter our behaviour towards them. We make our facial movements slower and more exaggerated to babies than we would towards adults. Broad smiles, cooing noises and pursed lips, huge eyes of mock surprise, eyebrows raised high on the forehead – we hold these and many more expressions for longer than in adult conversations.

The eyebrow flash, where we raise our brows quickly and briefly, is used between adults as a non-verbal 'hello'. It happens so quickly – one-sixth of a second – that it is almost invisible. But when we make the same expression to a baby, we do it much more slowly.

This special way of communicating exaggerated expressions with babies seems to be practised all over the world. Anthropologist Eibl Eibesfeldt has observed it in many distinct cultures, including among the Yanonami in South America, the bushmen of the Kalahari, and the Eipo in Papua New Guinea. So when a 'basic' expression is shown by a baby, the mother and other adults respond to it immediately. This is where the inbuilt expression begins to be modified from very early on.

We take, say, the natural inborn smile, and gradually help to instruct the infant by demonstrating all the consciously created variations. We know basic expressions are inbuilt, but we also know that they need to be elaborated, given context and nuance, even added to with new combinations of facial movement.

Opposite Babies are very
responsive to human faces and learn
early on to imitate expressions.

Our expressions then are rather like the colours in painting; there is a small number of primary colours that make strong, unambiguous 'statements', and there are endless shades of colour that can be created from them by mixing the palette. From the basic, inbuilt expressions we extrapolate to create variations that constitute a conscious, non-verbal language of facial expressions. It is in this rich interface between the basic and modified expressions that we can learn a lot about ourselves as individuals and as a species. The smile, for example, is an expression in which we can clearly see basic and consciously controlled expressions alternating, interacting, even competing.

THE SMILE: SHOWING AND HIDING FEELINGS

We all have a natural smile, the basic and universally recognized expression of happiness. When happy, we make this expression whether or not anyone else is around. Of course, in everyday life we also put on conscious smiles. These are smiles that are not natural expressions of happiness, but are for purposes of communication. They help to smooth the path of social interaction. Sometimes they are used to hide natural expressions, which would give away too much of how we are really feeling.

We might assume that the smile is physically straightforward. In essence, it is a stretching of the mouth around our teeth, with either the lips closed or open. We pull our lips back, and also turn them up at the sides. But there is much more to it than that. There are many sorts of smiles. Paul Ekman, long one of the world's experts on expressions, estimates that there are eighteen varieties.

The human smile is visible from a long distance, and may have evolved as a means to make friends and placate enemies.

The natural smile has a quality all of its own. If it is a genuine, spontaneous smile, then interesting things happen with our eyes, too. They sparkle, or soften, depending on the context and meaning of the smile, and the skin crinkles up around our eyes, forming what we call 'laugh lines'. This smile is very hard to fake.

Consciously created smiles are formed differently from the natural smile. In nineteenth-century France, biological researcher Duchenne de Boulogne

explored natural and conscious smiles in some classic and macabre experiments. He had discovered that by passing electrical currents through muscles, they could be stimulated into involuntary contraction. He found a homeless tramp in Paris whose face was partially paralyzed, and who agreed to be a guinea pig for the experiments. Duchenne pinned electrodes onto his subject, charged them up and recorded exactly how the twitching muscles distorted the face into 'expressions'. His findings are recorded in an extraordinary series of early photographs of the experiments.

Duchenne's major finding was that there were various different kinds of smile, only one of which was a true smile of enjoyment and pleasure. This is the smile activated by the zygomaticus muscles; in deference to this early expressions experimenter, it is now known as 'the Duchenne smile'.

The zygomaticus muscles, which run down from the eyes across the cheeks to the corners of the mouth, contract in a true smile of enjoyment – which is why the old adage about a sincere smile being visible in the eyes is true. In contrast, a consciously contrived smile involves only the risorius muscles and pulls the lips sideways, but not upward.

Another difficulty with trying to pass conscious expressions off as natural ones is that false expressions are slightly asymmetrical. This lop-sidedness occurs because conscious activity is controlled by the two cerebral hemispheres. One is more dominant, and sends a stronger signal. The face then responds in a slightly uneven manner. Involuntary expressions, in contrast, arise in the lower, unitary brain and affect both halves of the face equally. Asymmetrical expressions seem unnatural to us, and we tend not to trust a person who habitually uses conscious expressions where natural ones are more appropriate. We do not assume that the smiling used-car salesman is as genuinely happy to see us as he pretends.

There is nothing wrong with consciously contrived expressions, of course. We use them all the time, and they are essential in many social situations. They express how we feel and complement what we are saying. It is when

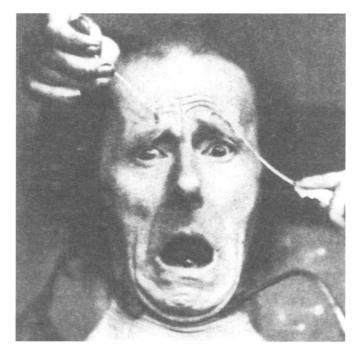

Opposite The natural smile… and the conscious one.

Above In the nineteenth century, biologist Duchenne du Boulogne discovered – through applying electrodes to the face – that the natural smile differed to the contrived one.

Above Dutch performance artist Arthur Elsinaar uses electrodes on his face to create his art.

we try to pass off a contrived expression as a natural one – to pretend that we are feeling something that we are not – that things begin go wrong. Also we use conscious expressions to hide our emotions, which otherwise would be visible through our basic expressions. This does not mean, however, that we are engaging in something as negative as 'deception' or 'lying'. We might want to mask our feelings to avoid hurting someone. And the process can be quite hard work, as anyone who has faked delight at receiving an unwanted gift can testify.

So in the normal rounds of social interaction, we know that people hide their thoughts and feelings sometimes, covering what would be a giveaway natural expression with a consciously constructed one to mask it. We try to make allowances for this, making our judgements of them by 'reading between the expressions'. But this is not easy to do. If it were, married couples would not need to consult the Love Lab.

Left Would you buy a used car from this man?

Below left Tracie Andrews, appealing for information on the 'road rage' murder of her fiancé in Worcestershire in 1996. She was later found guilty of the murder, and of fabricating the 'road rage' story.

Below Richard Nixon: would you buy a used car from *this* man?!

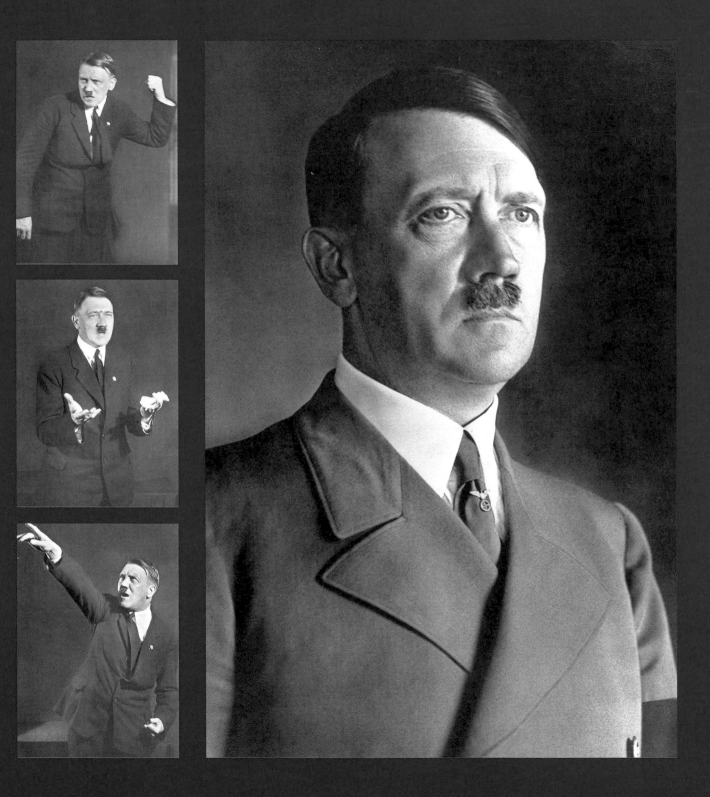

We make mistakes – sometimes important ones – because we misconstrue what a person's expressions are telling us. In September 1938, Prime Minister Neville Chamberlain flew back to Britain after his meeting with Adolf Hitler, to negotiate a peace treaty with Nazi Germany. Chamberlain wrote enthusiastically to his sister: '…in spite of the hardness and ruthlessness I thought I saw in his face, I got the impression that here was a man who could be relied upon when he had given his word'. Unfortunately, Chamberlain had misread Hitler, and a year later the two countries were at war.

But Hitler was a master performer. He made contrived, conscious expressions into a craft. He practised not only the words, but also the accompanying expressions for each of his key wartime speeches. Each statement was emphasized with an appropriate piece of non-verbal back up. He prepared for these speeches like an obsessive actor. He had each expression photographed so that he could judge its effectiveness and pull off a perfect performance. These photographs, still in existence, are a testament to how consciously controlled expressions can persuade people of one's convictions, especially if those people want to believe you.

We convince ourselves that we are good at detecting lies, but studies show that across the board, no matter what job we do, we are only about 50 per cent accurate at detecting a lie – which is no better than chance. Paul Ekman tested a variety of people working different jobs, under experimental conditions, to see if they could tell whether others (actors) were lying. The best individual human lie detectors were between 70 and 80 per cent accurate. They were secret service agents. So it is possible that practice does make you better at spotting a liar. The average scores for secret service agents was 64 per cent. Other professionals practised at judging people's expressions also scored better than chance: psychiatrists scored 58 per cent, judges 57 per cent, and police detectives 56 per cent. After that, subjects were close to chance.

The masking smile, a consciously constructed expression for hiding other emotions, is often used by public figures.

Opposite Adolf Hitler practised expressions and gestures for his speeches – and had himself photographed so that he could perfect them.

The natural smile and the masking smile use different muscles to produce a similar effect.

British Prime Minister Margaret Thatcher was a habitual user of a masking smile. When she answered questions from reporters, she often maintained a fixed and consciously constructed smile. Then, following the Falklands War, she appeared on a national television discussion programme where viewers could put their questions directly to her via a link-up. It provoked a celebrated incident when Mrs Thatcher famously lost her smile.

The programme took place some time after an incident in the Falklands, in which the British had attacked an Argentine warship, the *Belgrano*, despite the fact that it appeared to be sailing away from the British fleet and posed no threat.

Diana Gould, a member of the general public, appeared on the programme to ask Mrs Thatcher exactly why the order to torpedo the *Belgrano* was given. Mrs Thatcher claimed that the Argentine ship had been in British waters and was therefore a legitimate target. But the questioner was not satisfied and persevered with challenging questions. Mrs Thatcher had been forced into a position where she had to give an answer about where the ship was. Just before that happened her face betrayed her anger about being challenged in such a way. When the tape is freeze-framed, it can be clearly seen that Mrs Thatcher's fixed masking smile disappears for an instant, and is fleetingly replaced with a genuine expression, showing her fury. The expression lasted for less than one second, and was then immediately covered again with her permanent, fixed, masking smile.

The masking smile is frequently used to cover emotion. This sort of resolve, to prevent natural expressions from releasing our feelings, can become part of our cultural expectation, and referred to in national stereotypical terms, as when we speak of the English 'keeping a stiff upper lip', or the Japanese as being 'inscrutable'.

Paul Ekman studied such cultural variations in expression of feelings by comparing facial reactions of people to a film of material calculated to upset its viewers. Suction-aided birth, nasal surgery and an amputation would not be everyone's TV viewing choice! He asked Japanese and American students to watch this film, while secretly filming their responses. They all looked disgusted by the scenes, screwing up their faces and sometimes turning away from the screen. But as soon as the white-coated experimenters entered the

British Prime Minister Margaret Thatcher lost her 'masking smile' when under pressure during an interview – revealing, just for an instant, her true emotion: rage. Then she recovered her conscious smile.

Mr Kadokwa runs a smiling school in Tokyo. Young Japanese are traditionally taught that smiling masks emotions or indicates deference. By learning a more relaxed attitude to smiling, he hopes that his students can communicate better in the world business marketplace.

room, the Japanese students started to behave differently. They began to force smiles, and tried to look comfortable with the viewing matter.

Ekman calls this kind of altered behaviour 'display rules'. Each culture has developed its own set of rules about how much our face is allowed to give away, or conceal, our emotions.

The Japanese have a concept called 'wa' which, roughly translated, means 'propriety'. This results in strict rules governing all aspects of behaviour. In public, they try not to look one another in the eye, believing that it is rude, and strive to retain a neutral face, on which little emotion is expressed.

Right Young Japanese seem to be able to express their emotions more freely than the older generation.

When they do smile, the Japanese frequently use the more conscious smiles, such as the smile of deference. The muscular movements of this smile-face have their origin in the scream, and it is frequently indicative of anxiety. They also use masking smiles a lot in their everyday interactions, to cover natural expressions that they feel it would be improper to reveal. The Japanese, therefore, traditionally interpret the purpose of social smiling quite differently from people in the West.

The cultural differences in facial expressivity can confuse business negotiations. While the Japanese make much less eye contact, and are often silent for longer periods, American business people are used to associating gaze aversion and hesitation with deception. Now that the West is beginning to influence Japanese lifestyle in so many ways, some Japanese people are beginning to abandon the idea that the smile is basically always used to mask other emotions and, indeed, some of them are learning how to smile in a western fashion – to relax and allow natural smiles to occur more often.

What with cultures changing, and the fact that there are as many as 7,000 expressions, some of which we may be trying to mask part of the time, it is not surprising that we can sometimes misinterpret other people's expressions.

The rules of engagement

BIG BOYS DON'T CRY?

Conventions for 'displaying' emotion through expression are widespread across cultures, and so are perhaps based closely on biological imperatives. For example, in many cultures there appears to be an unwritten rule that men should show less emotion than women – and most certainly should not cry. This is the male archetype represented in western popular culture in recent decades by screen heroes such as John Wayne, Clint Eastwood and Bruce Willis, who portray characters untouched by emotions.

But things might be changing. After the Gulf War, General Norman Schwartzkopf appeared on Barbara Walters' television talk show in the United States. After a set of very personal questions he broke down in tears. The interviewer was surprised. 'Generals don't cry, generals don't get tears in their eyes.' But the general explained that he set

Teaching babies cultural variations on basic expressions has resulted in differences in greeting rituals around the world. Westerners kiss, Polynesians stroke each others' faces, Lapps and Mongolians rub noses, while Arabs kiss beards.

himself rules about when he was and wasn't allowed to cry. He would not cry in front of his men in the thick of military operations, he said, because then he was their leader and their figurehead. He would, however, cry in front of the same men in church at Christmas. He felt this was permissible, because in that situation he was acting as a surrogate father figure, and it was then appropriate for him to act as an emotional focus point.

Schwartzkopf is one of the most popular military leaders America has ever had. Perhaps his progressive attitude to crying will help to lift some of the restrictions of this 'unwritten rule'.

Political figures are careful to obey display rules, and fear disapproval from the public if they are seen to be transgressing them, by showing taboo emotions in inappropriate settings. President Clinton appeared to overstep the mark when he was caught on camera unawares, laughing in conversation with someone at a state funeral. He suddenly realized he was being filmed, assumed a mournful expression and made as if to brush aside a tear. But he was then criticized by the press and the public for insincerity — he was seen as trying to pass off a consciously constructed expression of sadness as a truly felt emotion.

Tears are made up of water, oils, proteins and salts. Chemical analysis of tears shows that some of these proteins have been associated with prolonged unhappiness when they build up in the body. Tears excrete these substances that make you unhappy. So we can finally prove scientifically why a good cry may make you feel better.

BEING EMBARRASSED

Display rules about crying might be changing in Britain too. The 'stiff upper lip' gave way at the funeral of Diana, Princess of Wales, when men as well as women wept openly. However, this was an exceptional occasion. Expectations might be shifting, but most men would still be embarrassed to be caught crying in public. We feel the emotion of embarrassment when we realize that we have been exposed to greater social attention because we have done something wrong or inappropriate. An embarrassed expression acts as an apology. Typically, we avert our gaze with the eyes moving downwards. We

Previous pages Boys learn early on that tears are a stigma – that crying is for 'girls' only.

Opposite A self-conscious Naomi Campbell after a fall on the catwalk.

might smile, briefly, and put our hand to our cheek. Sometimes we blush. The whole response takes about five seconds.

Embarrassment appears in children no earlier than about the age of eighteen months, much later than other expressions. It is at this age that a child becomes more fully aware of the people around and their sense of 'proper' behaviour. We can feel embarrassed by our actions only after we have learned, as children, the social rules of what is right and wrong, allowed and forbidden.

The expressions associated with embarrassment act as an apology. Studies carried out in courtrooms in the USA proved that defendants found guilty who blushed and looked embarrassed after the verdict was read out, received shorter sentences than those who appeared unrepentant.

YES AND NO

Of course, learning what behaviour is approved and disapproved is a major challenge for a young child. Much adult–child interaction in the early years focuses on signs of approval (Yes, with head nodding) and disapproval (No, with head shaking) – for 'your own good', of course!

Zoologist Desmond Morris suggests that head nodding and shaking may be connected to the movement of the infant when it is put to the breast. It can often be seen to move its head up and down in a searching movement to find the nipple. This up and down action means 'yes, I am hungry and want to be fed'. He suggests that it is this up-and-down head action that has survived into the adult non-verbal lexicon as the affirmative nod.

But when the baby has finished feeding, it rejects the breast in one of two ways: it either twists its head to one side or it raises it up. Both actions remove its mouth from the region of the nipple. As before, these actions have survived as adult gestures. The twisting of the head to the side becomes the sideways head shake meaning 'No'.

The head nod and head shake seem unambiguous, and therefore it should be simple to understand what people mean by them. In most parts of the world, this is true. But not everywhere. An exception is a region in Italy, south of Rome and north of Naples. If you speak to some Italians who live here,

they nod in order to agree, but there are others who may use another gesture: this is based not on the infant's sideways twist of the head but on its upwards tilt. This is because that area of Italy, in the region of the Massico mountain range, was the limit of the northern expansion of Greek influence over 2,000 years ago – and the Greeks traditionally used this upward head toss as a gesture of agreement.

Around the rest of the world, we rely on head nods, and many other small expressions, to set the rhythm of complex interactions with other people. When two people are talking, a single head nod from the listener indicates that they have understood what is being said, and wish the speaker to continue. Rapid and repeated head nods indicate that they wish to stop listening and take their turn to speak.

Above Eye contact and head nods are vital to both listener and speaker in maintaining a conversation.

The head nodding expression is therefore not only crucial to learning sanctioned behaviour, but also to our capacity to converse at all.

And we combine nodding with other subtle clues. When we are talking to someone, we usually start the sentence looking at them. Then, during the course of speaking, we look away from them – perhaps because we need to minimize visual distraction in order to formulate our speech. When we come to the conclusion of the sentence we look back again. The person listening to us may give us a very small nod and that, coupled with the fact that they are clearly listening to us, means several things – that they understood what we have said, have basically accepted or agreed with it, and are inviting us to continue. But if our conversational partner wants to cut in, their facial expression will warn us that the interaction is about to change. Perhaps they want us to explain something we have said. Or they might want to take their turn to speak.

We spend up to 75 per cent of conversation time looking at the person we are listening to, but only half of this amount of time when we are talking.

EYE-TO-EYE

In normal conversation, the periods of eye contact are very short. We glance up at one another for brief periods of about three seconds, but will hold one another's gaze for only a second or two – any longer makes the speaker and the looker feel nervous. Prolonged eye-to-eye contact of more than ten seconds indicates that one of two things are about to happen – the two people are preparing to fight, or make love!

Prolonged eye contact activates the nervous system, forcing up our heart rate and blood flow. It is an unconscious reaction. However, the

Below Some butterflies have wing markings that look like staring eyes, to frighten off predators.

confrontational impact of prolonged eye contact is strongest in cultures such as those in Britain and America where it is 'rude to stare', because staring is considered hostile. In the inner city ghettos of the United States, staring is considered very confrontational, readily leading to violence. To counteract this, some successful black rap musicians are being coached to help them to feel more comfortable with more frequent and longer-lasting eye contact, and the social smiles that fill the wider world in which they now move.

Previous pages Children sharing the same expression and, by inference, the same emotion – happiness.

Of all primates humans are the only ones with the sclera (white part of the eye) exposed – others have sclera that match the skin colour round the eyes. So it may be that the staring human eye carries even more of the expressive impact for the person seeing it than the more camouflaged-eye stare of other animals.

By contrast, in India or Greece it is quite normal for people to watch or stare at strangers in a way that westerners would find embarrassing. Greek people have said that they find the English rather rude, because when they visit London they feel ignored – no one stares at them.

Our nervous system registers a stare automatically. Whether we feel threatened, charmed or included by it is culturally determined.

Intimate eye contact stimulates the hypothalamus to send hormones

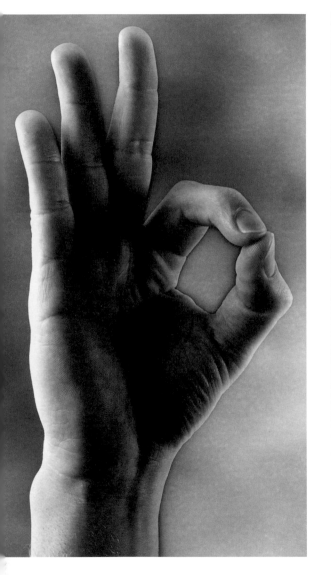

Above During the Gulf War,
the US Defence Department
warned its troops not to make their
familiar A-OK sign, since locals
would view it as a symbol of the
evil eye – a malevolent stare,
like the one on the right.

flooding into the system, creating a physiological reaction similar to panic. The body cannot differentiate between the excitement of eye-to-eye contact in aggressive and romantic situations. Fortunately, we can usually recognize the difference, and this is where sustained eye contact takes a turn for the better. In the right context, even a glance held a fraction longer than normal may be perceived as an act of intimacy. It can be done in sharing a private joke, making a point, or as an act of flirtation. The glance penetrates the private psychological space of the other, and also reveals one's own. If the glance is not 'sanctioned' by the other, their natural response is to glance away. But if the flirtatious glance is returned, and held, then you're in luck!

Flirting by both men and women often starts with a quick upward glance, followed by averting the gaze and then another bout of eye contact, and a friendly smile. People who are mutually attracted to one another will talk and move their heads close together, looking at each other face on. Eye contact goes up substantially. Not only do they nod frequently, the conversation begins to take on the form of a mirror dance – if one nods, the other will, if one blinks the other will and so on.

Chewing, crunching and grinding all act on the body's parasympathetic systems and calm the body down. This is why a meal is such a good idea for a date!

IN THE BLINK OF AN EYE

We saw in Chapter 1 that blinking evolved as a means of preventing our eyes drying out, moistening them with tears. In addition to this biological function, blinks also communicate. We are very sensitive, unconsciously, to cues gained from people's patterns of blinking. At one level they help to co-ordinate interpersonal communication. In conversation we blink to show others that we have grasped an idea.

Sometimes actors blink at the wrong moments, destroying the illusion that the words they are saying are spontaneous and reflect real thought processes. Politicians, too, often blink at inappropriate times when they are talking. We sense the unnaturalness, and often mistrust what they are saying.

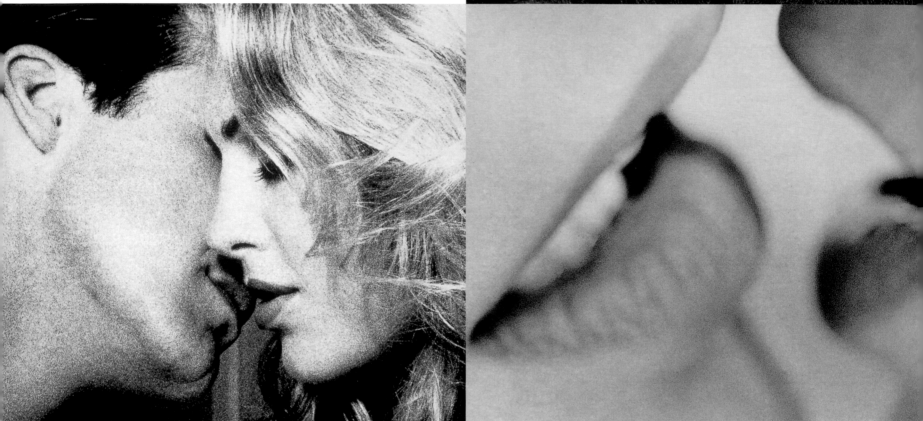

Above The kiss represents facial communication of the most intimate kind!

Top right Pygmy chimpanzees enjoy French kissing too!

How we feel also affects the way we blink. If we are absorbed in some task we blink very little. When concentration wavers, we release flurries of blinks. A bored child in a classroom will be blinking about eleven times a minute, compared with six times for a more assiduous classmate. Our pattern of blinking may reveal not only how interested we are in our activity – but also how interested we are in the person we are with.

THE LOVING KISS

If flirting, blinking and lust lead us anywhere, it will often include kissing. Exchanging intimacy by pressing your lips to someone else's is certainly face-to-face communication of the most pleasant kind. We do it because the lips are sensitive, and the sensation feels good – our heart rate goes up to over 100 beats per minute, and breathing rate and blood pressure also shoot up.

Kissing may have had its origins in something very simple: pursing our lips. As babies, we suckle at the mother's breast by pushing the lips forward, keeping them soft – and this facial movement survives into adult life as the loving kiss.

In feature films today we frequently see people kissing. But screen kisses have a turbulent and controversial history, starting in 1896 with a film called, appropriately, *The Kiss*. This four-minute film features two middle-aged actors, May Irwin and John C. Rice, sitting demurely on a sofa and then leaning in to kiss one another. One critic said it was 'absolutely disgusting'. Which brings us back to one of our most basic expressions.

DISGUSTED!

Disgust is an expression that has its origin in an ancient bodily action called the pharyngeal gag reflex. If we feed a child or an animal a bitter or sour substance, it will immediately push out its tongue and screw up its nose. It's a protective biological reaction when we think we are being poisoned. We spit out the food and block out the smell of it as quickly as possible.

Disgusted expressions are processed by the anterior insular section of the brain. This part of the brain is generally used for processing information about the stomach. So the biological origin of the disgust face (pushing away poisoned food) has been retained in the layout of the brain – even though the cause of the expression of disgust may now be a screen kiss!

So disgust is an emotion deeply rooted in our biological make-up. It seems to have extended to social and psychological aspects of our lives through the idea of contamination. We feel, for example, that if something 'disgusting' is put in a glass of water, and then removed, we still do not want

The 'Hays Code' regulated the film industry in the USA right up to the 1960s. One of its 'moral' requirements was that if two adults were seen kissing in a bedroom scene, at least one of them should have a foot on the floor at all times. This was in order to maintain a sense of propriety and avoid any suggestion that the kiss might be leading further!

to drink the water. We know it is not literally 'contaminated', but in our minds it has been 'violated' by the repugnant item. And so by the same route, we can feel that our sense of values is contaminated by behaviour. For example, if we witness cruel behaviour towards a defenceless animal, we might feel 'disgusted'. A basic emotion connected to rejecting spoiled food has ended up powering some of our most highly moral values.

Opposite The expression of disgust first appears in early infancy.

The virtual face

As we have seen, facial expressions are an extremely rich language. In a few minutes of interaction with a spouse, they may reveal sufficient information for an expert to predict the outcome of the marriage. A smile for a yearbook can indicate the relative success of our relationships over the next thirty years. Our brain assesses alarm and likely threat situations from other people's expressions. The face bonds babies to their mothers, and vice versa. Expressions connect people emotionally through the smile, and regulate complex social interactions through masking. They offer the liberation of Mr Kadokwa's smiling school in Tokyo. They afford us the opportunity to catch the British Prime Minister's anger, and the American President in a social gaffe. They give us the chance to release tension by having a good cry, to flirt, kiss and get disgusted.

And yet today, we are doing more and more faceless, and therefore expressionless, communicating. E-mail and other computer-aided connections can be extremely effective adjuncts to face-to-face interaction. But business is relying on it increasingly as a medium for complex communications and negotiations. And for the young, especially, much social time is spent in 'conversation' with people who are remote geographically, and whom they have never met personally. There are advantages to this, of course, in that people can begin to interact without the relationship being swamped initially by the sort of mutual physical attraction issues we consider in Chapters 4 and 5.

However, we need to be more clear about what we are losing. Expressions are a potent part of our connection with one another. What are the consequences of communicating without them?

In 1984 psychologists set up a study known as the 'map task', to assess how far facial signals aid communication. Couples were given a map and had to negotiate the safest route through a set of hazards. Some of the couples could see one another, whereas others could only speak to one another. The people who were able to see their partner's facial expressions were both more accurate in their instructions, and more sparing in their use of dialogue.

When in 1997 the same interactions were studied over video link, where eye contact was made possible through the arrangements of cameras, the people did not seem to behave in ways characteristic of normal face-to-face communication. One possible factor is that although people can see each other's faces in video link, much of our judgement of expressions is done with reference to the social and physical context in which they are being expressed.

Contagious expressions

Some expressions seem to thrive in the company of other people. Both yawning and laughter are 'catching', setting off others around us. If we see someone else yawn, read about yawning or even think about yawning, 55 per cent of us will ourselves yawn within five minutes. It seems to be contagious. Can you feel a yawn coming on now? Yawning appears to have some kind of social function. Although infants yawn right from birth, it only becomes contagious after the first year of life, perhaps when the child is becoming more socially aware. Scientists have little idea why these expressions became contagious in humans, but in no other animals.

As laughing is contagious, we prefer to be entertained in company, and that's why television situation comedy programmes usually have a laugh track – we hear other people laughing, and it makes us feel more like laughing ourselves (at least, that's the idea). There is no proof of how or why this happens. But hearing laughter can be a relief of tension for us, and gives us permission to express the laughter sound ourselves.

Both yawning and laughing need the company of others to induce their 'catching' effect. Virtual communication probably means fewer yawns – but, sadly, it probably means fewer laughs, too.

Opposite Yawning is the most contagious of expressions. If we see someone else yawning, we are very likely to be doing the same thing within a few minutes.

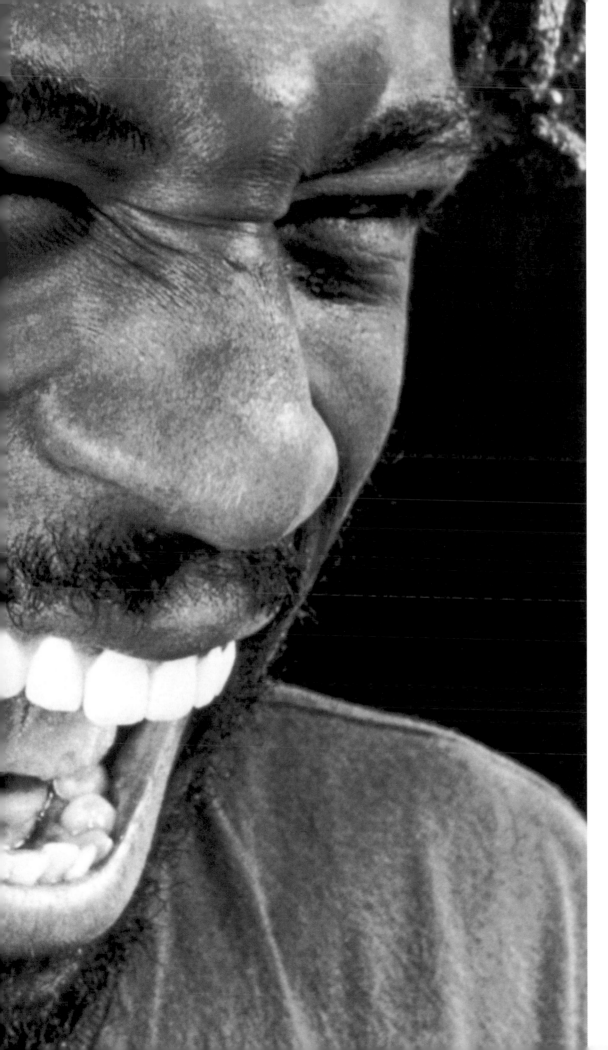

Left And laughing is very contagious too...

Right More contagious laughter.

Below Intimate eye-to-eye contact.

When people communicate in person, face to face, they each make a continuous stream of facial expressions. We normally think of these expressions as being there for their communication value, in letting each other know how we feel about the conversation. But there is now considerable research evidence to show that the making of facial expressions has a direct impact on how we feel. Called the facial feedback effect, this process means that expressions reinforce the emotion that gave rise to them, strengthening and making richer our world of feelings.

The facial feedback effect seems to work through physiological cues, in which the position of our facial muscles feeds information back to our brain. When the brain realizes we are smiling, it releases the hormonal response that usually accompanies our state of feeling happy. Our 'happy face' and 'feeling happy' works backwards as well as forwards. This means that even social smiling, consciously induced, can result in our feeling happier.

Face-to-face interaction stimulates facial expressions. Communicating with people by e-mail does not, for there are no expressions to which we need to respond.

Of course, it might be possible to make the computer itself responsive to our expressions, and such 'empathetic' computers are now in development.

Knowing that expressions can alter our mood, a Japanese scientist has built a facial robot that has the potential to create and respond to expressions. Her muscles are made of pistons and her skin is just silicone, but her camera eyes can spot our smile, and smile back at us. The scientist hopes to market her as the companion of the future – something to keep lonely people company, which might also keep them healthy.

As more and more of our communication is done electronically, the one thing we are missing out on is face-to-face contact. But not for long. The information technology industry is developing avatars – virtual faces – that in the future will speak to colleagues and negotiate for us in

Four hundred years ago, William Shakespeare was well aware of the facial feedback effect. In *Henry V*, the king motivates his soldiers for battle by commanding them to alter their facial expressions:

Then imitate the action of the tiger;
Stiffen the sinews, summon up the blood,
Disguise fair nature with hard-favoured rage;
Then lend the eye a terrible aspect;
Let it pry through the portage of the head
Like the brass cannon; let the brow o'erwhelm it
As fearfully as doth a galled rock
O'erhang and jutty his confounded base,
Swilled with the wild and wasteful ocean.
Now set the teeth and stretch the nostril wide,
Hold hard the breath and bend up every spirit
To his full height!

international meetings. People will open an e-mail and up will pop our avatar, speaking our words and responding to their comments with appropriate expressions.

So the honest among us will have their face digitized and all our faces will have the worldwide coverage of today's superstars. But of course our avatar does not have to look exactly like us. The temptation to straighten up the

Below Fumio Hara's robot, which smiles in response to your smile.

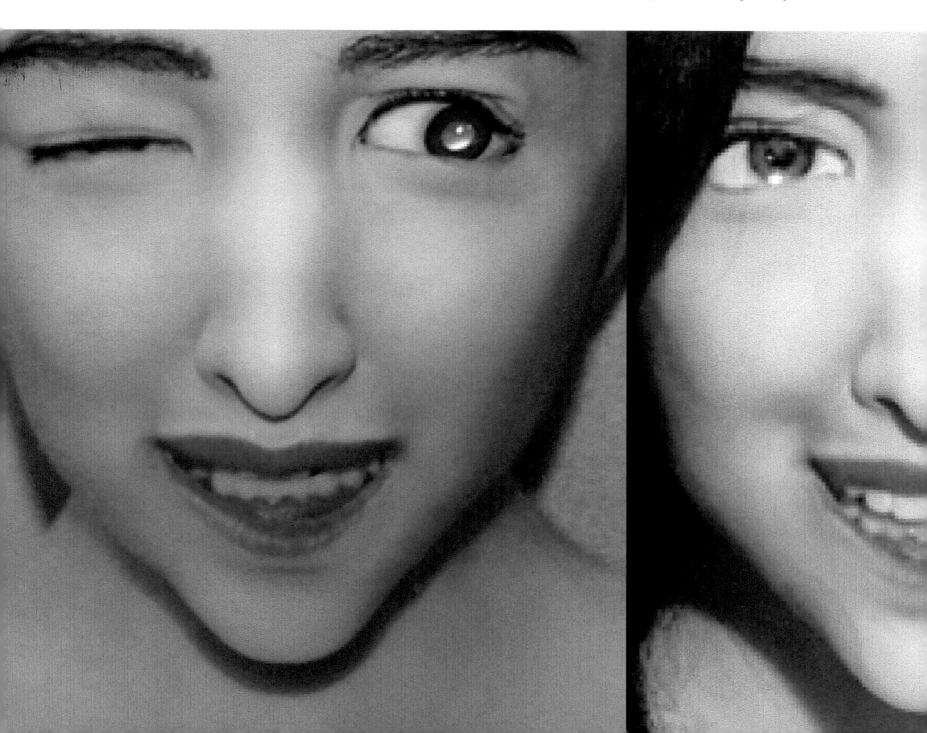

nose or even adopt a completely different appearance will be too much to resist for some of us. Will we be able to trust facial expressions at all in the future when we have made our virtual faces more beautiful than our biological ones? And how will we decide how to make a beautiful face? In the next chapter, we turn our attention from our ever-changing expressions, and consider what it is about a face that leads us to think it beautiful.

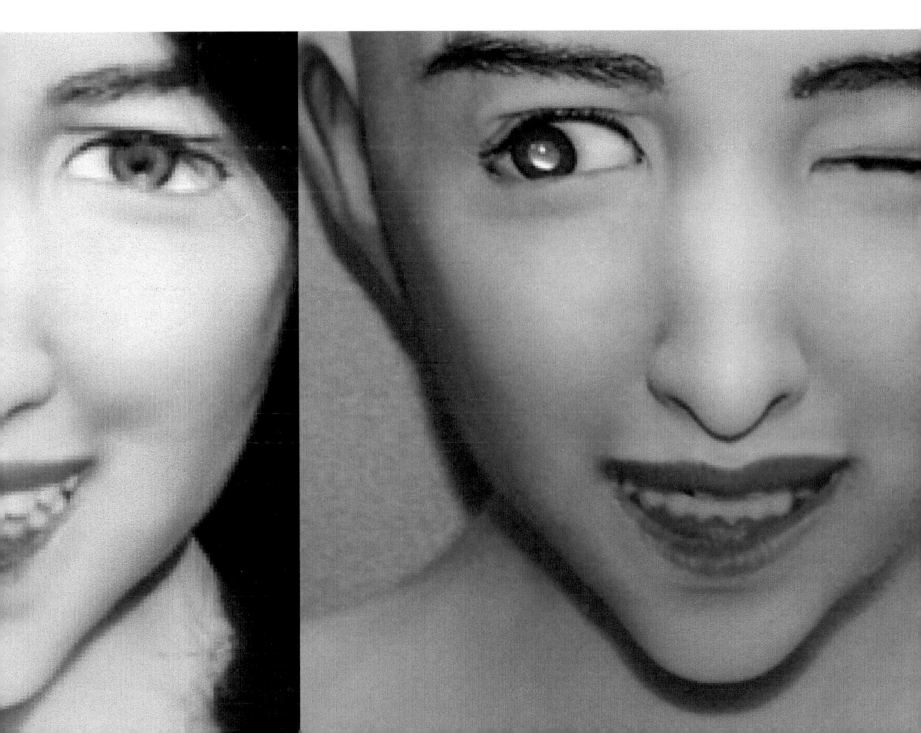

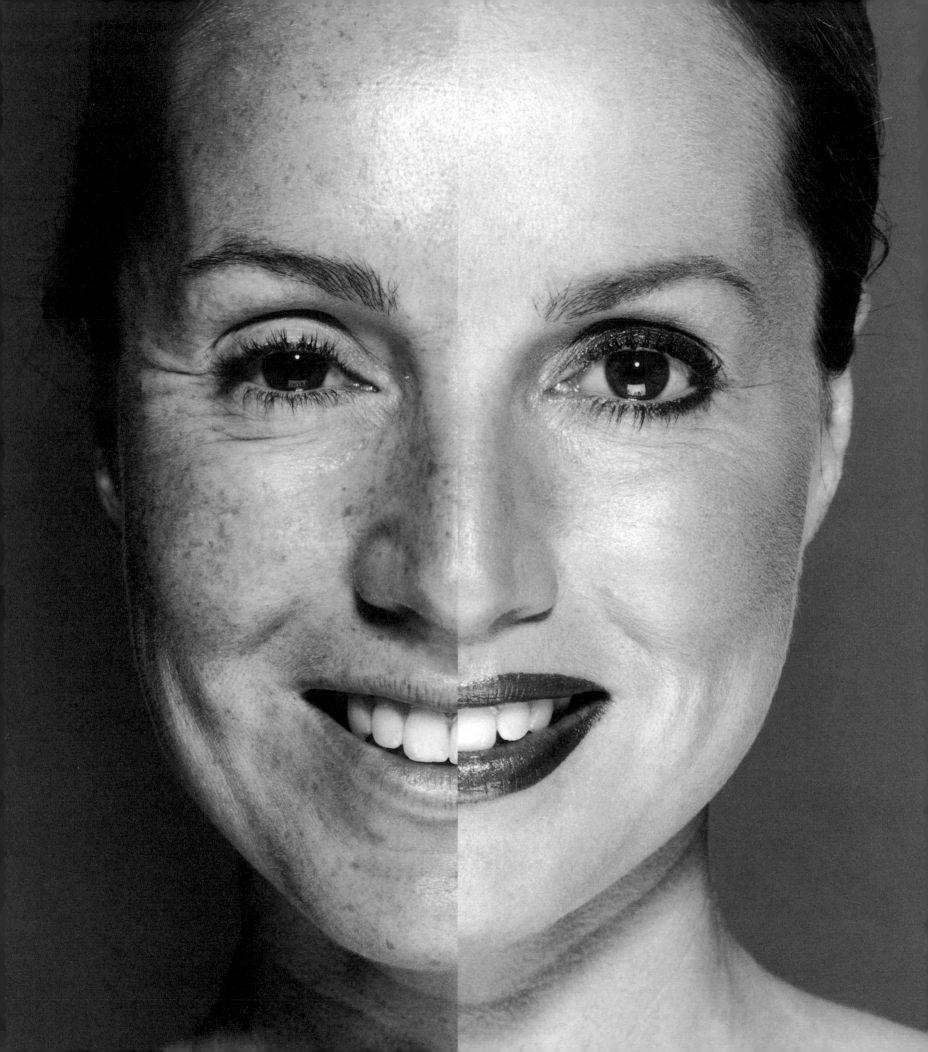

beauty

CHAPTER 4

'BEAUTY IS IN THE EYE OF THE BEHOLDER' – such a commonplace phrase that it has become a cliché. But do we believe it? Are our judgements of who is beautiful purely a matter of personal taste? The answer rests on a blend of Greek philosophy, evolutionary biology, mathematical formulae, babies, sex and personal chemistry.

In one way beauty is in the eye of the beholder because we shall always argue about whether Brad Pitt is more fanciable than Leonardo DiCaprio, or whether Julia Roberts is more fanciable than Anna Kournikova. We each have our personal tastes in the 'beauty icon'.

Left to right Some beautiful faces: actor George Clooney; model Iman; violinist Vanessa-Mae; and actor Brad Pitt.

But beauty is also not in the eye of the beholder: remarkably, all over the world, people seem to agree on which faces are considered beautiful. This is true whether they are judging men or women, and people from their own or different racial backgrounds.

In 1993 anthropologists Douglas Jones and Kim Hill journeyed to two relatively isolated tribes in South America, the Hiwi Indians of Venezuela and the Ache Indians of Paraguay, to explore this phenomenon. These tribespeople had seen very few people from outside their own village culture. They had no access to television, and so had not even seen images of people from elsewhere in the world. Jones and Hill showed them a range of photos of women from various cultures, and asked them to rate them on beauty. Their

All these faces are beautiful. But they are beautiful in different ways, and they each evoke within us quite different kinds of feelings. Sometimes it is awe, sometimes love, sometimes deep longing, sometimes lust, sometimes our hearts miss a beat... and so on. Each of these emotions is a response to a particular kind of beauty. Or, perhaps more accurately, to different aspects of the multi-faceted quality of beauty.

judgements were compared with ratings of the same photographs by people in Russia, Brazil and the United States.

People in all five countries were attracted to female faces with delicate jaws and chins, and large eyes. Other cross-cultural studies have confirmed

Opposite Julia Roberts.

Left Calista Flockhart.

Following pages Faces of today:
Kate Moss; Gwyneth Paltrow;
Anna Kournikova; Kate Winslet;
and an icon of an earlier decade,
Sophia Loren.

that women are regarded as beautiful if they have smooth skin, big eyes and plump lips. Research has also shown that men and women agree with each other on which faces of women are most beautiful. And when men rate the attractiveness of other men's faces, they not only show moderate agreement among themselves, but also agree with the ratings made by women.

On the other hand, beauty is in the eye of the beholder if we consider it across time. It may be that we generally agree on who is beautiful now, but facial fashions have changed in the western world over the last few decades. Recently many famous female models and actors have fought to lose weight, because the very thin face and body has come into vogue. So much so that the extreme thinness of some models, such as Kate Moss, caused an outcry, because it was believed to be setting an unhealthy example to young girls. Such thin icons of fashion seem to be suggesting 'not eating' as a way of gaining control and approval of female bodies – perhaps contributing to the increasing incidence of anorexia nervosa among adolescent girls.

Recently the editor of a leading fashion magazine declared that she would feature no more of the 'thin brigade' like Calista (*Ally McBeal*) Flockhart – and yet many magazines still find that the thin figures and faces sell more copies. There are of course fuller-faced icons of beauty like Kate Winslet, star of the film *Titanic*. But so far the thinnies still have it.

Yet if we go back to the 1970s and 1960s, facial beauty seemed to be a more natural size. And in the 1950s the fuller-faced and figured woman, such as Marilyn Monroe, was more popular. There may be other aspects of the beauty of these women that transcend the fashion trends of the decades, but certainly we can see changes.

The eighteenth-century philosopher David Hume argued that beauty 'exists merely in the mind which contemplates them: and each mind perceives a different beauty'.

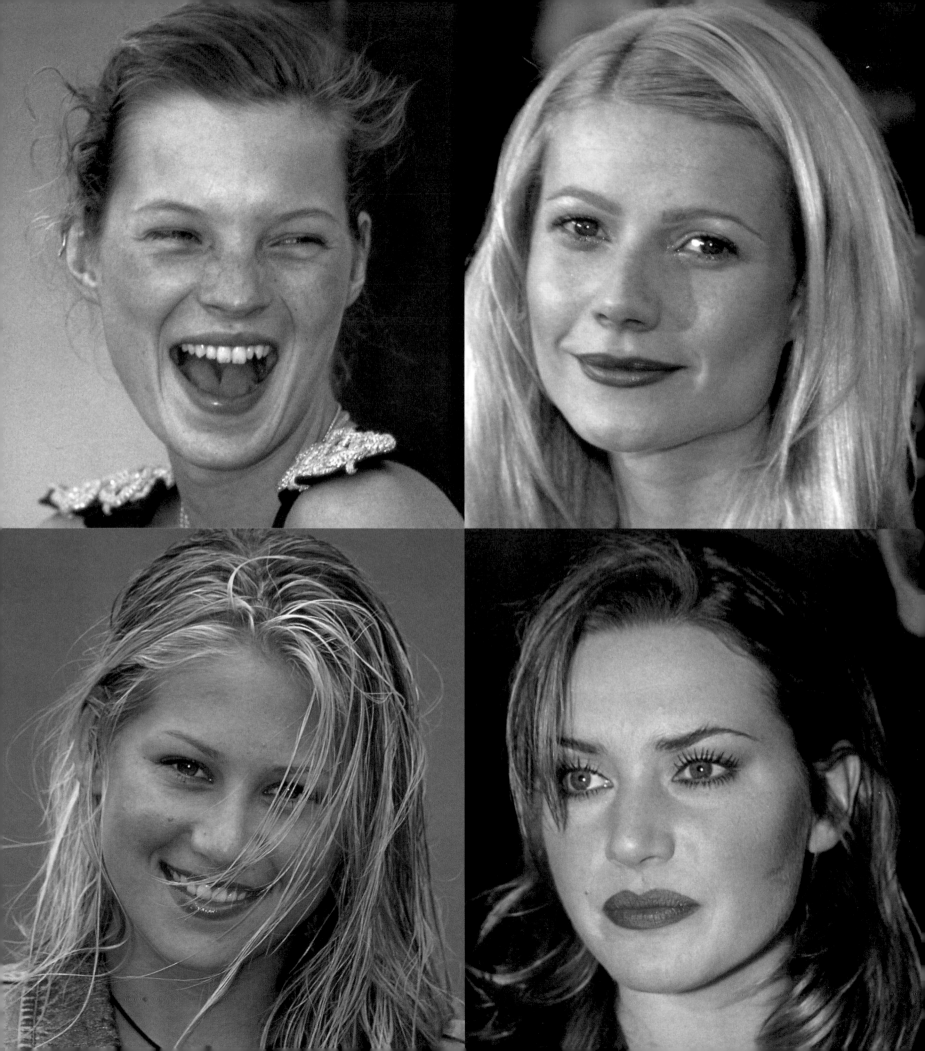

These fashion trends in beauty are easier to track in women than in men, because judgements of a man's beauty tend to focus on more disparate features. However, there has been a recent shift towards the youthful or even boyish looks of film stars such as Leonardo DiCaprio or Matt Damon, and away from the rugged features of Sylvester Stallone or Arnold Schwarzenegger. This change in 'facial fashion' may reflect a new cultural acceptance in the West of the more emotionally open 'new man' of the 1990s, whose sensitivity has been associated with more submissive facial features. If we go far enough back in history, however, we can see that attitudes towards beauty have always been changing.

The Greeks and Romans both prized eyebrows that met over the middle of the nose – a feature then considered a mark of great beauty. In the late 1600s double chins were considered fetching – of course, this was a fashion statement for the faces of the well-to-do only, who clearly could afford to eat well.

Fashions are often recycled after surprisingly long periods of time. In fourteenth-century England, the aristocratic 'girl about town' would not have looked out of place in the most fashionable nightclubs in modern-day London or New York: her eyebrows were completely plucked, her forehead shaven and the remaining hair dyed bright crocus yellow.

For both men and women, in the middle years of the twentieth century a deeply tanned face was fashionable, before it was realized that too much sun was a health risk to the skin. Suntans looked healthy, and were a clear way of showing that you had enough money and leisure time to take holidays. Facial fashion clearly reflects the status priorities and attitudes to health pertaining at the time.

Opposite Research shows that people considered beautiful – like Marilyn Monroe – are not necessarily happier than everyone else.

Left Three hundred years ago, a double chin was fashionable. George Hamilton demonstrates that in the 1970s it was a suntan.

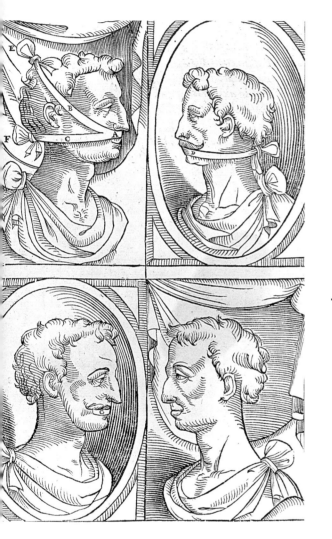

Above Tips on improving the profile from a 1597 book on surgery.

Opposite Supermodel Christy Turlington being 'made up'.

There is yet another view about where our judgements of beauty lie. Saying that beauty is in the eye of the beholder makes the whole process sound like a matter of the mind. It would be better to say that beauty is in the breast of the beholder, because beauty is often a matter of the heart. Sometimes our whole lives are transformed because we meet someone we find beautiful. However, our judgement of that person's beauty is almost certain to be much more inclusive than mere physical attractiveness. As we have already seen, the face reveals the emotional life of a person. When a large group of men and women were asked what characteristics they would like to show in a photograph, they mentioned beauty, but also intelligence, friendliness, kindness and 'being nice'. These are essential elements also of our views of beauty, and we shall consider them later in this chapter.

The essence of beauty

Over the past 2,000 years and more, people have sought to identify the defining principles that underlie all judgements of beauty. Searching for factors deeper than personal preferences, fashionable looks, or other more superficial dimensions, artists have endeavoured to calculate the perfect dimensions for a beautiful face. The ancient Greeks developed the notion of the Golden Mean. This dictates that the face is split horizontally into three equal sections: in a beautiful face, the brow should be one-third of the way down from the hairline, and the mouth one-third of the way up from the chin. Overall, the face should be two-thirds the width of its height.

Eventually the proportions were expressed as a mathematical formula that divided the perfect face into a ratio of 1:1.618, where the ratio between the smallest parts to the larger is the same as the ratio between the larger part and the whole face. This can sound arbitrary to us today. But it helps to remember that the Greeks, and artists of the later Medieval period, had notions of beauty that were far deeper than mere judgements of people's physical attractiveness. Human concerns were considered part and parcel of the secrets of life, knowledge possessed by the gods. It was felt that this Golden Mean numerical association was so perfect and seen in so many places in the natural world, that it came close to the divine. It was like an attempt to find an

equivalent for beauty of Einstein's $E=mc^2$ formula for time and space. Thus, statues depicting gods and goddesses were meant to represent this universal formula of beauty.

Of course, it is these proportions that are responsible for the 'identikit' faces of thousands of statues and paintings across history. And yet even today many cosmetic surgeons still use these measurements as a guide for their work.

Opposite American Cindy Jackson has spent thousands of dollars on plastic surgery to make herself fit her own perceived notion of beauty.

The search for the essence of beauty by formula continued right through the centuries. Artists' vision of the beautiful face evolved into the notion that it should conform to proportions that divided it not so much into thirds as into sevenths. Botticelli's Venus fits into sevenths: the hair comprised the top seventh, the forehead the next two sevenths, and the nose a further two sevenths. Another seventh was taken up by the space between nose and mouth, and from there to the chin occupied the final seventh.

As with the Greeks, these artists were depending not only on their observations of the human face, but also on deeper notions of how the world is constructed. They believed that there was a principle whereby the whole

Above Elizabeth Hurley measuring up to Botticelli's Venus.

world conformed to analysis into sevenths. This so-called 'Golden Section' could be applied to the beauty of anything in nature or in art – for example, it was used to evaluate the beauty of a landscape as well as a face.

However, these formulae do not match up today. The features of modern models do not conform to the Golden Mean. In one study, those judged as having the most beautiful faces had eyes much further apart than the width of the nose, as advised by Dürer and da Vinci. So this proportional measure of beauty does not capture the essence of our contemporary judgements. Perhaps models are chosen today according to facial features that are different from those of classical beauty. Nonetheless, some of their portraits are undeniably beautiful to our eyes. The faces have a transcendent beauty, because so much of early western art was religious. Painters such as Botticelli and da Vinci were trying to capture the Beauty of the Spirit.

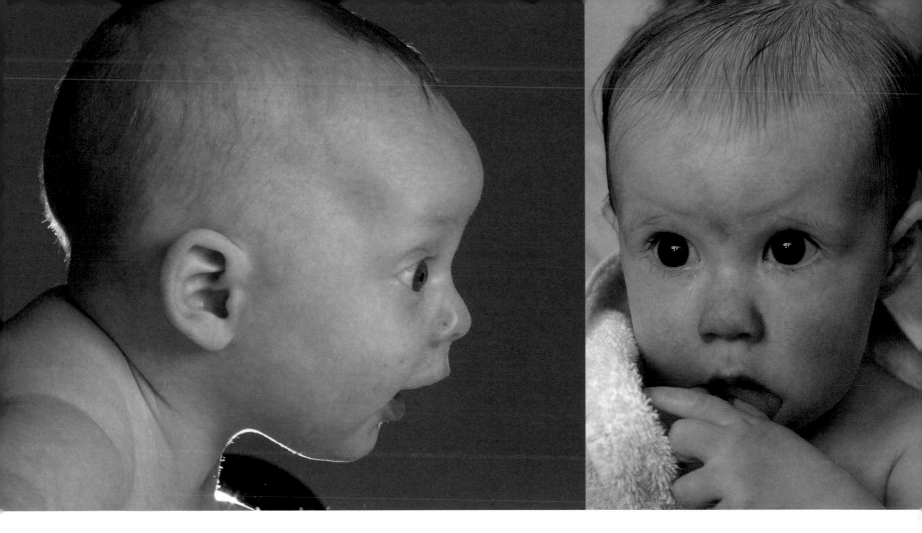

Baby face beauty

Today, however, we are more inclined to biological and psychological analyses of our world. One such approach to understanding judgements of beauty concerns our almost universal, probably inbuilt, reactions to the faces of babies.

Babies have a special kind of beauty. For many of us, tender feelings of protectiveness well up when we look at an infant's features. We find them 'cute' and we have the desire to cuddle the baby and look after it. The hallmarks of baby 'cuddliness' are chubby cheeks; a high, prominent forehead in relation to the small face (the infant's face is only one-eighth the size of the head, whereas an adult's is one half); relatively large eyes; smooth skin; a small mouth for sucking; and a head that is large in proportion to the body.

It seems that in the course of evolution, these features of babies' faces have developed to evoke emotional responses from us that protect the child. It makes sense that this should be so, of course, for small babies are so vulnerable that the species would not have survived were we not programmed to look after them.

Above and opposite Beautiful babies – the big eyes, chubby cheeks and large head provoke feelings of protectiveness in adults, vital for the survival of the human species.

141

In fact, our responses to the special beauty of a baby are so effective and universal that they have been exploited to maximize sales of commercial products. 'Cuddly' toys are often given attributes of the baby's face, and bring out the 'aaahhh' sound we make when taken with sweet, vulnerable beauty!

In general we are all suckers for this baby-face formula, as Walt Disney noticed many years ago. Think of Bambi. The large-eyed, soft-faced little deer orphaned by a forest fire brought lumps to the throats of people of all ages. And in his early films the most famous animated creature of them all, Mickey Mouse, appeared with proportions that were closer to a real mouse. But gradually his eyes were drawn in increasingly large proportions, an exaggerated version of a baby's face, and by the time Disney had finished they occupied 47 per cent of the area of his face.

The more a baby's features conform to this 'ideal', the better they seem to be in triggering the adult's desire to love and protect. Psychologists have observed that mothers with babies who have more pronounced infant faces spent more time looking at them, and cuddling them, than those mothers of less attractive babies.

Something we like about beauty in babies is carried into our judgements of beauty in adults. Certain people retain their babyish features into maturity, and our responses towards them suggest that they still evoke some of the same reactions. People with large eyes and rounded cheeks are more likely to be smiled at, told intimate secrets and cuddled by those closest to them. But they are also more likely to have been further into their teens before having their first sexual experiences. So their young features extended this period of being treated in a babyish manner.

Men are attracted to some babyish features in women's faces. Women's noses are proportionately smaller, wider and more concave than men's. Their mouths are smaller than in men, and the upper lip is often shorter. The jaw and brow ridges are less pronounced and the eyes look bigger. And these are the same features that distinguish a child's face from an adult's.

Opposite Baby features bring out the protective instincts in all primates.

Below Colonel Oliver North faced severe charges over illicit sales of US military equipment to Iran. His boyish features may have contributed to the thousands of supportive messages he received.

Below Japanese manga animation
films depict baby-faced young adults.
The popular Barbie doll also has
babyfaced features.

Opposite The US child glamour
pageant, in which young girls are
dressed precociously as adults.

But perhaps the clearest evidence that baby-like features enhance our judgements of beauty in women comes from a study by anthropologist Douglas Jones, who fed the images of various magazine cover girls – with their large eyes, small noses and plump lips – into a computer that correlated the size and proportions of people's faces to their age. It estimated the models to be between six and seven years old! Their faces did not really look that young of course, because there were other facial cues to their adulthood, but it is revealing that the facial measurements of these desirable and beautiful women would be more typically found in younger girls.

What, in that case, can we make of glamour pageants for children? These contests of looks and talent are very popular especially in the United States. They consist of pretty young children, mostly girls, between the ages of three and young teenage years, being dressed and made-up like glamorous adults. Much of the atmosphere of these pageants, fiercely competitive as they are, is to mimic in a precocious way the adult world of the parents – who do all the work of training and preparing their youngsters.

These contests do seem to enter a kind of shadowy, borderline world in which the innocence of childhood is mixed with the high-octane fuel of sexually attractive beauty. There are visual cues with babies and young children that restrict the way we respond to them. But if those cues are suspended, as when dressing them provocatively as attractive young adults in beauty pageants, then we can feel the tension created by it.

So in some ways, the beauty of a baby is no different from the beauty we respond to in an adult. But in other ways it is very different, because baby-faced proportions release our feelings of protectiveness. They are also extremely important to children, because they will normally guarantee that a beautiful baby does not stimulate in us a sexual response that is totally inappropriate to the age of the child.

What the beauty of babies certainly shows, then, is that we seem to have an inborn response to beauty, including automatic 'cut-off' mechanisms to protect children. Are there other clues to how our response to beauty is biologically ingrained in us?

The attraction of symmetry

One of the factors preoccupying artists over the centuries was that beautiful faces seemed to be symmetrical. This seems to be corroborated in modern research. On the whole, we find symmetrical faces more attractive. This preference starts early, and may be inbuilt. Psychologist Judith Langlois collected slides of human faces and had them ranked for attractiveness by adults. When she showed the slides to infants three and six months old, she found that they stared longer at the faces the adults had rated most attractive.

Opposite Miss World, the ultimate beauty pageant for women.

Below Women's earlobes become more evenly matched in size at their time of ovulation.

But there is another aspect of beauty that was most definitely not present in the renaissance religious icons: sexual attraction.

In the 1980s scientists discovered that humans give off pheromones 'aromas' that are undetectable but which we take into our olfactory organs. The term 'pheromones' comes from the Greek for 'to excite'. And they do alter our behaviour. These non-detectable aromas play a part in sexual attraction in all sorts of animals.

Biologists Randy Thornhill and Karl Grammer tested the effect of pheromones on humans. They asked a number of men to sleep in the same T-shirt for three nights, without the aid

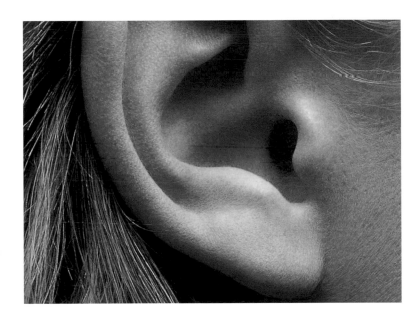

of deodorant. The shirts took on the pheromones of the men. Meanwhile they measured the men's faces for symmetry, and asked a group of women to assess the attractiveness of each man's face, based on its degree of symmetry.

After the three nights, another group of women were given the T-shirts. They had no idea which of the men had worn which T-shirt. They were asked to smell each shirt and give it an attractiveness rating! The surprising result was that the most symmetrical men were also the owners of the most attractive 'smells'. But only when the women were ovulating – in other words, when their bodies were primed for sex – because that is when they were ripe to conceive. The rest of the time the T-shirts were rated as uniformly horrible!

What this indicates is that sexual attractiveness has a strong... (no, not smell) biological component. The symmetry of a man's face is reflected in his pheromones which, although not detectable as an actual aroma, is taken in and processed by the brain. It is tempting to regard this as a relic from our ancestors, who were perhaps able to detect a potential mate just by sniffing them.

And the symmetry seems to work both ways. Biologist John Manning measured a number of features of women's bodies, including the size of their ears, and the third, fourth and fifth fingers of each hand. He repeated these measures periodically, over some months, with thirty healthy women between the ages of nineteen and forty-four. The results confirmed that the symmetry of their body increased by 30 per cent in the twenty-four hours prior to ovulation (confirmed by a pelvic ultrasound). In other words, their ears became closer together in size, as did the corresponding fingers of each hand. So women are most symmetrical at the time they are most likely to conceive.

So why should symmetry be related to sexual attractiveness? Evolutionary theorists suggest that a high degree of symmetry may be an indication of particularly good genes, and perhaps resistance to the sorts of disease that can cause asymmetrical development. Highly symmetrical animals attracting other symmetrical mates would increase the hardiness of the species. Certainly mate preferences in other animals are consistent with this argument. Birds and insects favour symmetry in their choice of mate. Female swallows prefer male swallows with tails that are both long and symmetrical. Female zebra finches prefer males with symmetrical coloured leg bands.

And in research studies, men and women with highly symmetrical bodies were asked about their sex lives. They reported having had a greater number of sexual partners, and starting intercourse at a younger age, when compared with people rated as equal in facial attractiveness but with less symmetrical bodies.

Opposite Men and women with more symmetrical bodies report having more frequent sexual partners than average.

Above Glowing skin and sparkling eyes are a testimony to good health.

Opposite Film actress Angelina Jolie, renowned for her voluptuous lips.

Women seem to be particularly sensitive to the symmetrical qualities of men just at the time they are ready to conceive. This would mean that over thousands of years, natural selection has favoured people with those good looks, and has inbuilt into our brains the tendency to like and admire such looks, and to want to mate with people who look like that. If we have babies with people who have symmetrical bodies and faces, we will have more reproductive success.

This is intriguing evidence for the biological nature of our sexual attraction to each other. However, it needs to be treated with caution when applying to humans. For example, it is possible to produce an image of a perfectly symmetrical face on a computer, by taking one side of someone's face, reversing the image and joining it up with the other half, and thus making one half of the face become the 'whole face'. But people looking at these faces tend to find them eerie or strange, and not as attractive as we would expect. And recent studies at the University of Massachusetts have demonstrated that when people were shown photos of individuals ranging in age from late adolescence to late adulthood, relatively attractive people were mistakenly rated as healthier than their peers. Their attractiveness rating did not, in fact, correlate with their medically assessed health.

What else, apart from symmetry, tells us that a potential partner has 'good genes'? Are there other 'beauty' indicators to draw us to such people? Right across the animal world, flashy plumage, crests and colour displays advertise the genetic quality of their bearers. And it's not so different in humans.

The human equivalent of the peacock's tail is the face. The face is like an advertising hoarding that seems to provide the potential mate with information about health, fertility and appropriateness as a sexual partner.

For a man, the ideal female mate to reproduce with is someone who is young, strong and healthy, but above all, fertile. For a woman, the ideal mate is a fertile man with high-quality sperm to produce strong children. He must also be able to provide materially for their upbringing.

These characteristics translate directly into face terms. The youthful female face is particularly easy to read. She has large eyes and plump lips. The apparent size of these features peaks in the early twenties when fertility also peaks. As fertility declines, the lips become drier and shrink, the skin begins

to wrinkle, and the flesh round the eyes causes them to look smaller.

The ideal male face has a strong jawline, nose and brows, characteristics that emerge at puberty through the action of the hormone testosterone.

A beautiful face is a fertile face, and our obsession with beauty is less to do with aesthetics and more to do with sex than we ever imagined. However, there are young people who look healthy, vigorous and fertile, but are not beautiful. Clearly, vitality and fecundity are not enough to define beauty. So let us turn the enquiry back around from fertility to beauty, and see what other features we are attracted by.

Men's choice of lovers

Most people, when asked to morph photos of faces on a computer to achieve a face they like, tend to go for an averaging of the possibilities. Faces formed this way tend to have regular features, and to be attractive.

But is this the same thing as beauty? If we now add to these average faces those characteristics that we find particularly desirable in each sex, an even more attractive face emerges. For example, if a 'morphed' female face has the eyes made a little bigger the face immediately looks more beautiful. And if the characteristics are increased by an additional 10 ten per cent, the female face, as it gets more beautiful, seems to get more and more youthful. While we like the attractiveness of averaged faces, we find beautiful those faces that have exaggerated 'attractive' features.

But then, if we continue exaggerating the female characteristics, the face begins to become much more overtly sexy. The features become provocative. It is what is known as a 'glamour' pic. The eyes are so large as to be sexually inviting. The lips are plumped up enough to represent sexual arousal.

These are the sorts of faces we see in glamour models and in erotic poses, in pornography. Women with faces that look like this, or who have applied make-up to emphasize such features, are presented purely as sexual objects. These exaggerated faces, while not necessarily the sort of faces men are 'comfortable' with for longer-term relationships, are definitely the sort of faces that inspire lust. And in looking for 'mates' this is what many men go for, at least to start with, because they think it will get them sex.

Below Most men choose life partners of average attractiveness, even though they are excited by an overtly sexual appearance.

Opposite Hugh Hefner, a lifetime of proving he is a high-ranking, dominant male.

We sometimes think of women choosing men who will be reliable, stable and responsible fathers. But men choose women as long-term partners on the same basis. It is the latter with whom they have (planned) babies, and they want women who will stay around to help raise the babies. So although men are excited by sexually explicit looks, that is not necessarily how partners are chosen.

What they are looking for in those circumstances are women who have nice looks but much more to them. An average face, with a few added female characteristics, seems to be about the perfect mixture. Despite what the biology lesson tells us.

Now the women get to choose

If it is the women who are doing the choosing, what kind of male beauty are they attracted to? We have seen that they often have other factors in mind when choosing a male partner, but certainly they do have feelings about which men they fancy: 'tall, dark and handsome…'.

While women can definitely be turned on by a sexually attractive man, they tend to choose their partners in a different way from men. Anthropologist John Marshall Townsend showed people pictures of men and women who ranged from beautiful to below average, and who were described as training to be either waiters, teachers or doctors (low, medium and high earning professions). They were asked to indicate in response to each photo whether this was a person they would like to chat to, date, have sex with or marry. The women always preferred the best-looking man with the most money!

Yvette Hoyle, who runs a dating agency in London, says that men always select dates on the basis of looks. However, men can be intimidated by women who look too glamorous, because they feel pressured to 'perform'. Which is why they are sometimes more comfortable with women of a pleasant-looking but more 'average-attractive' appearance — in the long term.

But when this was not an option they would go for an average-looking doctor, or even an unattractive doctor, as readily as a good-looking teacher. For women, status compensated for looks. What underlies this attitude? A clue comes from the research on women's preferences for different kinds of male face.

We noted early in this chapter that male 'facial fashion' has recently shifted from heavily masculinized features, over to a softer, slightly more feminized appearance. At St Andrews University, psychologists David Perrett and Ian Penton-Voak invited women to sit before a computer screen, presenting a morphed male face – an attractive average face created digitally by the computer. She could make it look more or less masculine by moving a bar from left to right. Pushing the bar to the far right created a subtly masculinized face, and to the left a feminized face. Each woman was tested on a variety of different faces at different times in her menstrual cycle. After long tests in Scotland, Japan and South Africa, these psychologists have discovered that the faces a woman classes as most attractive are quite feminized men. But then, only for those few days in her cycle when she is actively fertile and would be able to conceive, the face she classes as most attractive is a heavily masculinized one.

Above Women tend to choose an average-looking, or even unattractive, professional over a good-looking manual worker.

Opposite The actress Barbara Windsor and her younger husband

155

Big jaws and noses, in the style of the 'Marlboro Man' advertisements, mean that the man has plenty of testosterone. As we have seen, in evolutionary history this meant that he would be likely to provide healthy offspring, and would also be physically strong enough to protect them from the dangers of the outside world. But there are disadvantages to having lots of testosterone. It makes men more aggressive and more likely to seek out new sexual partners – not likely to be a trait favoured by women of any era. Women may like men to look dominant, but not overly so, for it could indicate a lack of caring qualities. And the more feminine-looking man may be more caring and affectionate – they have higher levels of a substance called cortisol which damps down the aggressive aspects caused by testosterone.

So whereas men are predictable in what they look for in a pretty face, women are rather more discerning. They cannot reproduce all the time, being functionally fertile for only a few days each month, and so their choice varies. They are more concerned about long-term commitment than solely the quality of the genes. They want to find a mate who will stay, and take care of them and their children. In modern terms this translates into money, power and influence. This presumably is why, in 'lonely hearts' columns in newspapers and magazines, men put in information about their wealth,

whereas women do not. Typically, male correspondents mention whether they are financially secure, are car and home owners, while women concentrate on their attractiveness and 'fun' personality.

Left to right The rugged looks of Russell Crowe and Bruce Willis and the more feminized features of Leonardo DiCaprio and Jonathan Rhys Meyers.

Beauty as personal chemistry

Most of the experiments we have considered so far, intriguing and provocative as they are, are highly artificial. People are only looking at photographs – two-dimensional, bloodless representations of people. What happens when we meet a real live human being? Other factors colour our judgements of attractiveness.

Yvette Hoyle says that at her London dating agency, people who have lively and animated faces are invariably popular, because they are fun to be with. A mobile, fascinating face may be more attractive in 'real life' than one that conforms to traditional notions of beauty but is relatively lacking in expressiveness.

Evolutionary psychologist David Buss published in 1989 the results of a study of interviews with over 10,000 people from thirty-seven countries around the world about their mating preferences. He discovered that the

Above The archetypal lean-jawed cowboy of the famous Marlboro cigarette commercials.

most desired traits in a mate were not necessarily physical appearance – which was ranked only fifth by men and seventh by women – but kindness and intelligence. And a recent study by Susan Sprecher at the University of Illinois, in which people rated factors that attracted them to their partners, listed the two most important as warmth and kindness. This confirms what we know from experience, that in everyday life we are often attracted to people by factors other than judgements of physical beauty.

This subtlety in partner choice is colourfully illustrated by the Wodaabe, a tribal group who live nomadically in central Niger, in Africa, between the Sahara Desert and the grasslands. They are about 45,000 people, migrating with their herds of cattle, goats, camels and sheep. For nine months of the year hardly a drop of rain falls. For the remaining three months, rain returns. And at the end of the rainy season, and the greening of the pastureland, a magnificent celebration called the Geerewol takes place. For seven days, up to a thousand men participate in a series of dance competitions judged solely by women. During this week, women single out the most desirable men, choosing husbands and lovers.

In one of the dances, the young men vie to be judged the most beautiful. They dress uniformly, so that the attention is on their faces. They are resplendent in make-ups that accentuate their beauty – whiteness of teeth, largeness of eyes. They paint their faces red in the main Geerewol ceremony, and yellow in the other ceremonies. They also darken their lips and outline their eyes in black-purple and paint white lines over the length of their noses. The men look gorgeous and rather feminine. The choice of men for the Wodaabe women seems then, at first, to echo the studies we saw earlier, where women were preferring the computer-generated male faces that had rather more feminine features. But a crucial feature of the Wodaabe tradition is that in choosing their men, the women look in their faces and at their dancing for qualities of kindness, intelligence, warmth and charm. In other words, there is more to beauty than physical attractiveness.

We used to think that intelligence was a homogeneous faculty. But now scientists measure intelligence in many different ways, including emotional abilities, because there are so many complex aspects to it. A person can be 'high' on one aspect and not on another; we each have strengths and weaknesses.

And so it is with beauty – it is a multi-faceted quality. It is clear that there are many different ways of being physically beautiful, from having an expressive face to being a carved stone goddess, from being a protected beautiful baby to being a lust-inducing siren, and from being a feminized dreamboat like Leonardo DiCaprio to being a sporty hunk with a sweaty T-shirt. But in real-life choices of mate, issues of physical symmetry or social status are balanced by attention to kindness, intelligence and charm. Personal preference is not reducible to a formula based just on physical beauty. From this perspective, where we each choose what sort of beauty we are looking for, then beauty truly is in the eye of the beholder after all. Not that we have the confidence to believe this, generally. Much of the worldwide cosmetic industry is based on the notion that increasing our attractiveness to others enhances our lives – so let us now consider vanity.

Below A Wodaabe tribesman makes up his face in preparation for the Geerewol ceremony, where women choose their lovers.

van|ity

CHAPTER 5

Above Most of us cannot resist the chance to glance at our reflection in a shop window or mirror.

Opposite Actor Yul Brynner making up before a performance.

Actors know their faces. The lines, bumps and wrinkles are studied in brightly lit mirrors. Bone structure and skin texture are understood and used. Features are enhanced, disguised, changed and channelled by make-up. This is not vanity, but a necessary aspect of their craft. They use their faces as the leading edge of the characters they play.

So do we. Every day is like a small performance. The character is ourselves, and our faces reveal, and sometimes hide, the person inside. We are vulnerable creatures. We often find it hard to live with ourselves as we are. Fantasies of an enhanced self are common features of our daydreams. Many of us see our faces not as reflections of who we are, but as surfaces to be altered to conform to who we would like to be. Hairstyles are changed to enhance and flatter. Beards are grown to add interest, maturity and to conceal nature's 'errors'. Make-up is applied to imitate famous models, fashions and styles. In fact, awareness of our image for motives of social and sexual attraction is an accepted fact of life, for we believe that our appearance can make a significant difference to how others regard us.

Most of this is harmless, of course. Grooming is big business for the cosmetic industry, and slips into our lives virtually unquestioned. But there is a darker side. To be vain is to be overly obsessed with self-appearance, or particularly pleased with ourselves because of some imagined petty distinction. That distinction can be so petty as to mean 'normal'. We can be so afraid of being considered not normal, or average, or one of the herd, that we take action against others in order to make ourselves feel elevated. Being vain means comparing ourselves favourably with others. The dark side of vanity is prejudice and discrimination.

Imagine a world without mirrors

The mirror seems to be the essential ingredient for our indulgence of vanity. When the Wicked Queen demanded, 'Mirror, mirror on the wall, who is the fairest of them all?' she did not expect equivocation. She flattered herself that she was the most beautiful in the land, and preening in the mirror merely served to support her vain notion. When a clear rival to her self-delusion – her stepdaughter Snow White – eclipsed her in looks (and perhaps symmetry…) the Queen ordered her killed. Her interrogations of her newly true mirror reduced her to rage with its repeated assertions that Snow White was still the fairest in the land – and therefore alive!

Our daily encounters with our bathroom mirror may be less fraught, but the inspection of our features is haunted by some of the same insecurities. How do we look? Better than yesterday? Better than her or him? Whatever our delight or despair in gazing into the mirror, a clear reflection is something we take for granted.

This was not always the case. The ancient Egyptian and Greek aristocracy had fine polished discs of metal in which they could see the reflection of their faces. But the images in everyday mirrors were not so clear, and mirror-gazing was not the constant pastime of today. In the Greek myth of Narcissus, the young man was so unused to the notion of seeing his own face that he glanced into a pool and fell in love with his own reflection. Perhaps the easy availability of cheap, clear glass mirrors lies behind (or more accurately, before) our present-day absorption with our own faces.

Because male baldness is a result of an excess of sex hormones (androgens) that leads to loss of hair on the head, the only certain way to avoid baldness is to be castrated before puberty – this is a tribal ritual most men would be keen to avoid!

Above Young woman with a mirror, by Giovanni Bellini, 1515.

Previous pages Narcissus falls in love with his own reflection.

The first 'true' mirrors were made in 1460 by the Venetians, who worked out how to create clear glass. The aristocracy who could afford to buy these mirrors so enjoyed seeing themselves in full clear reflection that mirror-making quickly turned into a lucrative business. The craftsmen and merchants of Venice busied themselves on product research and development. In 1507 two Venetian brothers, Andrea and Domenico del Gallo, uncovered the secret of how to create the kinds of mirrors that led to the modern mirror we have today – a flat glass plate backed with a tin or silver amalgam. It provided a perfect reflection.

The brothers del Gallo 'possessed the secret of making good and perfect mirrors of crystalline glass, a precious and singular thing unknown to the whole world' and their process of manufacture was kept secret for more than 150 years. Everyone who could afford it wanted one of these new mirrors. They were exported to the wealthy countries of the world, and Venice maintained its stranglehold on the trade by any means possible.

By 1664 the purchase of mirrors from the Venetians was depleting the coffers of the vain French aristocracy. Numerous diplomats and spies were sent to wheedle out the secrets of mirror-making. Louis XIV's ambassador to Venice, Gonzi, even hired spies and eventually smuggled two mirror-makers away from the Venetian island of Murano and back to France. But mirror-makers who attempted to work for the French were followed by Venetian assassins and poisoned. The murders and intrigue continued until the French finally broke the monopoly, and discovered the secret of how to make the mirrors. By 1672 Louis XIV had installed the Galerie des Glaces at the Palace of Versailles – a whole hall full of mirrors.

It was to prove an early and prescient step towards the huge beautification industry today – on which we spend billions on cosmetics, hairdressing, dentistry and plastic surgery. The story of the mirror serves as a vivid metaphor for our absorption with our appearance. The invention of a clearer mirror led to an industrial secret worth protecting for a century and a half; and when the process was revealed more widely, mirrors were hung in their dozens, scores, even hundreds, in the great houses of Europe. Once we could see ourselves really clearly, there was no end to it. Like Narcissus, we fell in love with our own image.

Below Queen Ahmose Meryetamun, Egyptian Queen *circa* 1525–1504 BC.

Revealing disguises

Walk along any street in any city today, and you will be surrounded by people's subtle and not-so-subtle efforts to dress their faces – an assortment of make-ups, hairdos, moustaches, sunglasses, piercings and tattoos. We pay attention to the way we look to flatter our physical attractiveness, and to present an image that reflects who we think we really are – or, perhaps more honestly, who we would like to be.

THE STORY OF MAKE-UP

It certainly is big business. Nowadays the beauty industry is worth $45 billion worldwide, with global multinational companies dedicated to colouring, smoothing and decorating the face. Is this vanity gone overboard? Are we entering a new era of total self-absorption?

It could be argued that we are simply carrying on an ancient tradition. Make-up has been a favourite way to define oneself for a very long time. As long as 30,000 years ago, in Africa, our ancestors chose to decorate their faces with red ochre, a naturally occurring ore that they mixed with animal fats and

Right Renaissance ladies shaved their
foreheads to effect a fashionable
'steepling brow'.

Opposite For much of history it was
considered impolite to smile in
one's portraits.

Inset Mary Queen of Scots, judged
to be a great beauty in her time.

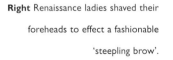

rolled into sticks. And the ancient Egyptians used iron and copper ores to decorate their faces. Eyes were the focal point, partly because of mystical beliefs associated with the power of the eye. But this also had a modern-sounding health aspect: it was worn to protect the eyes against the sun.

The make-up kit of the Egyptian noblewoman was elaborate. Egg white was used for facials; pumice stones and razors removed hair; kohl was used round the eyes; and various kinds of dyes decorated the eyes and lips. She even went into her tomb prepared with two colours of lipstick – all ready for a fashionable afterlife. However, favoured colours were scarcely those of today: a dark, purpley-black was the preferred shade for the Egyptian lady.

However outlandish face decoration is today, it cannot be said to be breaking new ground in bad taste. In Elizabethan times, people of wealth and leisure went to great lengths to make an impression. Women covered the face and neck in a thin paste made of white lead, water and vinegar – the forerunner of modern foundations. Pale, translucent skin was so prized that women were often bled to produce a wan appearance, and painted blue veins on to their faces to suggest fragility.

Of course we now know that white lead is highly poisonous. It is absorbed through the skin and cannot be excreted from the body. Hundreds of women in the fifteenth to eighteenth centuries died for the sake of this artifice. In seventeenth-century Italy a Signora Toffana was tried and executed for a series of manslaughters due to her special skin-whitening preparation – though it was not the women who used the product who suffered, but their husbands. The liquid was made with arsenic. It was absorbed faster through the mouth than the skin, and it poisoned more than 600 husbands who ingested it after kissing their wives' pretty faces.

While these white, alluring faces were slowly killing women and men alike, another problem arose. There was an outcry among men against women's personal adornment. Not because it was considered vain, but because the gentle art of seduction by appearance was considered to be deceit – make-up could mask the 'true' face. In 1770 in England a law was drafted that permitted divorce on the basis of this 'deception':

That all women … whether virgins, maids or widows, that shall
from and after such Act, impose upon, seduce and betray into

matrimony, any of his Majesty's subjects by the scents, paints, cosmetic washes: artificial teeth, false hair, Spanish wool [a form of rouge pad] iron stays, hoops, high-heeled shoes and bolstered hips, shall incur the penalty of the law now in force against witchcraft and like misdemeanours, and that the marriage upon conviction shall be null and void.

The law lapsed because it was unenforceable.

And it has not slowed the fascination with make-up as self-enhancement. Today motivational researchers employed to discover how best to market cosmetic products have discovered that many women are reluctant to think of themselves as covering up with make-up. Vanity still has a bad name. In the contemporary commercial world, it has therefore been rebranded as 'health awareness'. Advertising implies that the purpose of these products is to 'liberate' the latent beauty within, and reveal the most of our natural good looks. The illusion, at least, of naturalness is the goal. Make-up is no longer about lipsticks and rouges, but rather moisturizers and herbal face packs.

Nevertheless, among the young in many cultures, there is also a return to more showy, explicitly sensual make-up and dress, in which sexual display is upfront. Female pop stars strut their stuff in this style. Make-up aims to widen the eyes and plump up the lips – eyeliner and shadow, gloss and lipstick create the illusion of larger features. Lipstick also mimics the appearance of sexual excitement – our lips redden in association with sexual arousal and lipstick suggests that its wearer is in some kind of heightened erotic state.

The fashion world of eighteenth-century French and British high society was outrageously extravagant. The women piled their hair on top of their head so high, that it was said at the time: 'A lady's face often appears where one would expect her navel to be.' To achieve this effect, grease was rubbed into the hair to make it 'stay', and the locks lifted up around a frame of linen and wire, then stiffened and whitened with powder or flour. The ladies' scalps, clogged with so much organic matter, became an ideal breeding ground for vermin. When the hair arrangement was finally dismantled, the scalp had to be picked clean of lice and their eggs.

Opposite Portrait of an eighteenth-century noblewoman with hair piled high.

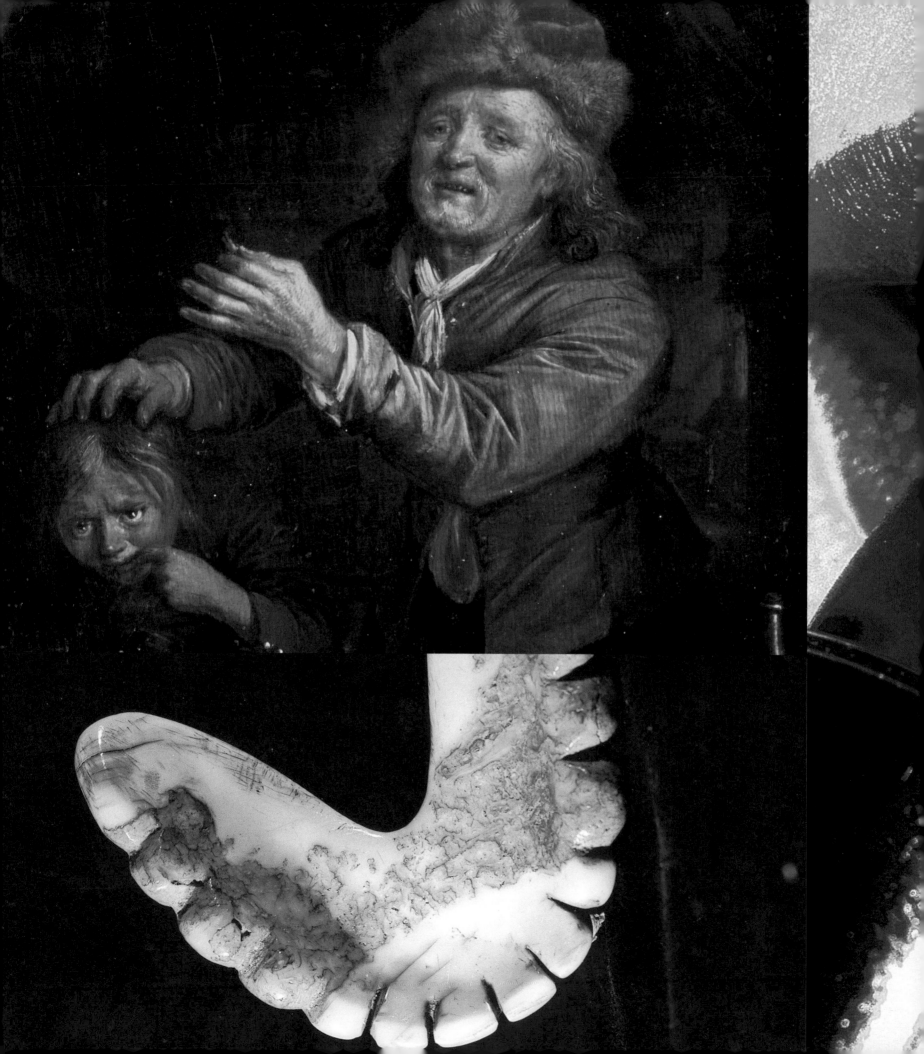

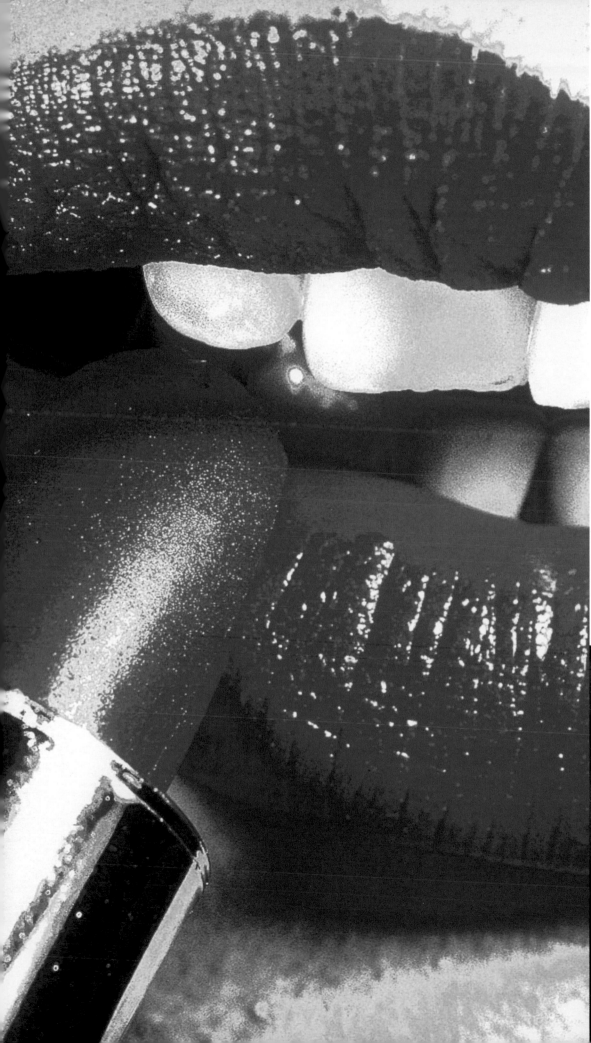

Opposite Tooth extraction (above) was a fairground 'attraction' in the 16th and 17th centuries. False teeth (below) first became used widely in the 17th century – made of porcelain, ivory tusk and … human teeth taken from corpses.

Left Lipstick enhances the redness of the lips, suggestive of a state of sexual excitement. Braces (below), for straightening the teeth, are common today.

LOSING OUR LOOKS

Age is the enemy of facial beauty for all of us. The hair loses its colour and becomes thin, the skin sags and becomes drier. The chin doubles, hearing and eyesight fade, earlobes grow bigger, and the nose broadens and lengthens. Wrinkles seem to be a major concern to many, for they make us look older – and research has shown that people's impression of our age influences their judgement of our personal qualities and capabilities.

However, wrinkles are not due purely to chronological age. Most dermatologists say that the majority of wrinkles are due to sunlight. Some claim that 'natural' age wrinkles need not occur until you're in your sixties or seventies.

Many beauty treatments on the market today are for maintaining the vitality of the skin, or for colouring greying hair. Facelifts, eyelifts, and hair transplants are among the more ambitious of the anti-ageing interventions.

MAKE-OVERS

But vanity also has a positive dimension. Sometimes we feel that our image is locking us into an identity that is limiting. We have spent so long living up to what is expected of us at home and at school, in the office and in our personal lives, that we have forgotten who is the person under all of it, sidelined, hidden, controlled.

We decide on a 'make-over', to allow our appearance to reflect our inner self more fully. But it is when we seek to change our appearance, that we glimpse another side of vanity. A relatively stable appearance renders us predictable, identifiable, and 'safe', and it is surprising how many people like us to stay that way. Very few changes are allowed. We can gain weight (but frowned upon), lose weight (approved up to a point), adopt a new make-up (sanctioned if not too radical). It is a sign of how important these external markers are that people often comment upon even gradual changes in appearance. A new hairstyle attracts reactions from friends and colleagues.

Wrinkles develop because there are two major layers to the skin. The upper, the epidermis, is self-renewing. The lower, the dermis, is the base on which the epidermis sits. These two layers are tied together with collagen, a naturally occurring elastic, that allows both layers to stretch and move with the muscles and then snap back into shape. When the collagen breaks down, wrinkles start to form.

Left The young Grace Kelly with a triangular-shaped face from the eyebrows and below.

Opposite Grace Kelly ageing beautifully – her face now more square-shaped.

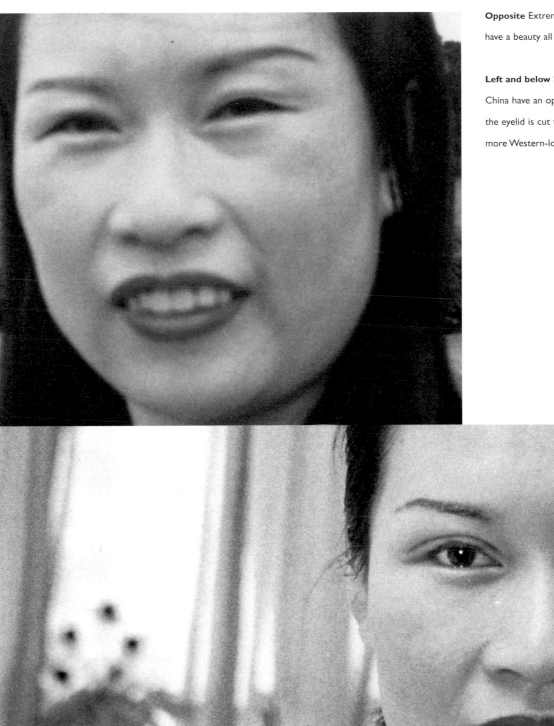

Opposite Extremely wrinkled faces have a beauty all of their own.

Left and below Some women in China have an operation in which the eyelid is cut to create a fuller, more Western-looking eyelid.

Left to right A variety of Madonna incarnations: the original 'lucky star' of the early 1980s; blonde ambition later that decade; and a short crop in the early 1990s.

Most comments are evaluative: the new hairstyle 'suits you'; the new beard 'makes you look scruffy'.

But occasionally reactions are even stronger. Altering our appearance can be threatening to people, because it suggests that we have somehow changed as a person. We have become an unknown quantity, a transformed individual, 'someone else'.

And so, while we are expected to be concerned with making our appearance 'attractive', in most subcultures a person who changes their image often, looking radically different on separate occasions and at various times, is considered 'shallow', weak, lacking a strong centre and therefore untrustworthy. Unless, that is, the changes are done with total conviction, from the inside. And few people have been as successful at this as Madonna.

Madonna is the greatest female image chameleon of all. She sprang onto the pop scene in the 1980s and since then her hair and make-up styles have changed more than any other celebrity's. She has copied the styles of most female icons of the century, from Marilyn Monroe to Greta Garbo and Eva Perón. She is one of the most photographed people in the world. Her photos show how she has adopted and subverted the stereotypes – she's been the dominatrix, the whorish vamp, the earth mother, the erotic temptress, the geisha, and most recently plain ol' mom, having another baby.

Madonna is least successful when people suspect that a new image is not 'really her' – like for example in her very erotic period, when the provocative images were criticized as being 'just for effect and publicity'. When Madonna's chameleon changes work best, it is because she presents a fresh image, but one that still seems to reflect aspects of her inner self.

Persona

Today, where cosmetics and other grooming products are so widely available, we seem to be identifying more closely than ever with our personal image. It means that our sense of who we are becomes almost synonymous with our facial presence. And the corollary is that if people do not recognize us, it is not just our face that is not memorable – it's our entire personality! This is why we are so agitated when people cannot recall having met us.

The term 'persona' originally referred to the masks that actors used in the classical period of Greek history. Implied in each mask was a complete personality, often of a deity, which possessed the actor.

But today, 'persona' is a psychological term, developed by the Swiss psychiatrist Carl Jung, and refers to the outward face of our psyche, the face that the world sees. And as we all know, what we show the world does not necessarily correspond entirely with our inner self.

Each of us has a number of personas, which we present to the world in the various settings we enter: work, home, social occasions and so on. Associated with our persona is an image that we display. For some of us, this looks the same no matter what we are doing. For others, it may vary considerably from one setting to another.

People may glimpse only a small aspect of us through our image, a mere sliver of who we really are. But if that is all they have to go on, then they have to extrapolate. People read a whole 'personality' into our persona. Because of this we can be very concerned with getting our image 'right' – otherwise, goodness knows what people will think of us!

We can be embarrassed when we meet someone from one part of our lives, in another setting entirely. The different personas we present reveal different aspects of ourselves. If the two settings are 'tea with the vicar' and 'all-night clubbing', it is likely that the two sides of ourselves on show are quite distinct. So vanity involves giving careful consideration to how we dress our facial persona. We want the image we project to be 'read into' in the best way possible.

Above Medusa, from the Greek myths, with her hair of snakes.

Opposite Men's hair fashions.

Tribal markers

Today, hairstylists are expected to achieve more than merely a flattering appearance. Teenage and young adults, more openly tribal than older people, use their hair to mark out affiliation with a particular subculture. In the West, this includes the current fashion for very close cutting or even shaving of male heads. This is vanity of a sort, but a major motivating factor in adopting this appearance is fear of being an outsider, of not belonging to a group. In this way, vanity is a restrictive force, playing more on people's weaknesses than celebrating their individuality.

Sometimes more extreme measures are called for in identifying our appearance with a social group. Ear piercing, for example, has been popular around the world for a very long time. Because it is relatively painless to have the earlobes pierced, it is a common expression of vanity today. Piercing of facial areas, and other parts of the body, which do register pain is a more extreme measure. It has a long history, and reaches severe forms in some cultures. Saucer-shaped discs inserted into the lips are found among the Botocudo in eastern Brazil, the Surma in Africa and the Thlinkits in North West Canada. The nineteenth-century explorer and missionary David Livingstone reported that in Africa the discs could reach up to 29 inches (over 70 centimetres) in size, restricting the wearer's speech and the ability to eat.

Opposite Women's hair fashions.

Below Samson's strength was in his hair – and he was reduced to a mere human when his hair was cut.

Today, in the West, much more modest facial piercing of the upper ears, noses, lips, and eyebrows has become fashionable for both sexes. Arguably it began as a 'tribal' marker, In Britain, for example, it identified people with facial piercings as belonging to the 'punk' youth culture of the 1970s. The unusual and seemingly extreme nature of metal in the skin, along with the initiatory nature of undergoing painful procedures, marked out these young people as 'different'. Like many initially unusual and exclusive fashions in clothes and music, it has gradually spread more widely, and today facial piercing is common among young people as personal decoration.

Clockwise from right Western ear and facial piercings; facial tattooing is traditional in French Polynesia; the Paduary, in Thailand, regard an extended neck as a feature of beauty; the Dogon of Mali in West Africa sometimes remove a diagonal half of each incisor, creating a triangle.

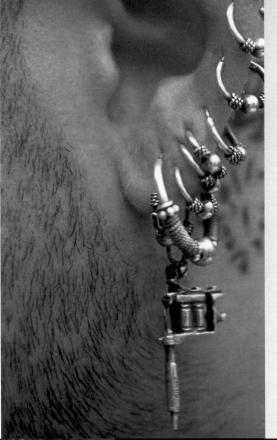

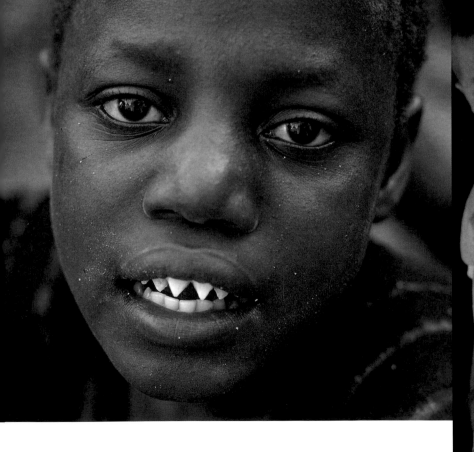

Similarly, in contemporary culture, tattoos on men were seen in the West as markers of tribal affiliation into a tough, macho subculture. Among women and now men in its new trendy guise, less extreme and more discreet tattooing is undertaken as attractive body decoration. However, tattooing has an ancient tradition. The Maoris travelled to New Zealand about a thousand years ago. A tribal warrior race, their society is very hierarchical, with eight ranks. Accession to a higher rank was marked by a tattoo, often on the face. The skin was pierced using hollowed-out bone chisels through which an inky mixture of soot and oil was poured. This produced the trademark black spiral patterns known as 'moko'. These beautiful spiral patterns were like a code, and initiates could read the family history, age and rank of a person from the moko.

The dark side of vanity

It is little wonder that vanity is more than a pastime – the pressures on us to look good are enormous. Studies in many cultures show that people considered attractive receive favourable treatment. It starts early. As we saw in the previous chapter, babies with characteristically 'cute' features receive more attention from their parents. And the bias continues into the school years. In the 1970s, teachers in 400 classrooms in Missouri were given the report card of a ten-year-old student, and asked to make judgements about the child's abilities, social skills and so on. The card detailed many aspects of the child's work, including grades, evaluation of attitude, work habits and attendance. But the researchers attached to the report card various photographs purporting to be the child – an attractive or unattractive girl or boy. Although the card contained a lot of information about the child's performance, the teacher's judgements were heavily skewed by the photographs. Teachers expected the good-looking children to be more intelligent, sociable, and more popular with their peers.

Good-looking children are not, of course, necessarily more intelligent than their plainer peers. Further research confirms that good-looking students often do get better grades where testing depends at least partially on teachers' subjective judgements. When grades are based solely on standardized tests, the advantage disappears.

Opposite Scarification of a person with dark skin leaves scars that reveal the brightly coloured flesh beneath. This is a Fulani woman in Nigeria.

Similar studies have also been carried out with adults. American studies of discrimination in employment have featured the placing of photos of attractive or unattractive job applicants onto identical resumés, which are then evaluated for employment. Even though the resumés are identical, attractive applicants are perceived as more qualified, and they receive starting salary recommendations as much as 20 per cent higher than the less attractive applicants.

Since people judged to be more attractive are treated preferentially, it suggests that concern with our appearance may be partly justified, and a matter of self-defence. Our experience, even if we are not aware of it consciously, tells us that if we look good life will be easier for us. Vanity has its value.

But simply reversing our reading of these studies shows us that vanity has its dangers. Just as attractive people receive favourable treatment, so unattractive people receive prejudicial treatment. In the above studies, unattractive children were judged, purely on their looks, to be less intelligent, less socially adept and less popular. They were given lower grades when teachers judged their work subjectively, even though they scored as well as good-looking children on objective tests. And as adults, less attractive people received starting salary recommendations as much as a whopping 20 per cent less than their good-looking peers with whom they were equally qualified.

Research is now beginning to examine directly the effects not just of looking 'plain', but of actual disfigurement. Psychologists Sarah Stevenage and Yolanda McKay set up a mock recruitment study at Southampton University. They found that while 100% of the interviewers hired an applicant with a normal appearance, only 55% did when the applicant had a facial disfigurement – despite the fact that they had identical resumés.

Being less attractive can also mean that you are considered to be of lower moral character. Jennifer Ramsey researched this dimension in kindergartens in America. She read the class a story in which there is one good character and one bad one. She then showed the children some possible pictures of the

Above Michael Crawford playing the title role in the musical *The Phantom of the Opera*.

characters. Children usually identified the attractive-looking person as the good character. Three- and four-year-old children assume that plain-looking people are morally inferior to good-looking ones. This discrimination continues into adulthood. Plain people are found guilty more frequently in court than attractive people. And American research has shown that in social settings, people are generally more aggressive towards unattractive people.

We used to say quite openly that people could be 'as ugly as sin'. Just as we favourably judge 'beautiful people' on the basis of their appearance, we also stereotype plainer people the same way. This tradition, of believing that a person's character could be read from their face, carried on well into recent times.

The most famous physiognomist was Johann Casper Lavater in the eighteenth century. This Swiss clergyman wrote a huge treatise on physiognomy, the subtitle of which was 'for the promotion of the Knowledge and Love of mankind'. He gave a clear description of the equation between beauty and goodness: 'The morally best, the most beautiful, the Morally Worst, the most deformed.'

Lavater's views have been totally discredited scientifically, but some of that deep-seated prejudice still lurks in contemporary attitudes. It is the dark side of vanity.

THE THREAT OF THE DIFFERENT

Enhancing and showing off our best features seems harmless enough. Natural. Even fun. But there is also an element of fear. Vanity is also about competition – does she look better than I do? Or a need for belonging – does he look different from me? It leads to bullying and favouritism among schoolchildren and judgemental, negative attitudes towards others perceived as 'different' in adult life. Today's obsessive self-regard and concern with appearance preys on our weaknesses. And inevitably, this negative aspect of vanity does produce victims.

For example, our attitudes to fat people echo much of the research above. They are often regarded, in the health-conscious, keep-fit regimes of today's vanity-culture, as dissolute. Overweight people are assumed to lack discipline. They are indulgent, eating whatever and whenever they want. This is

Above British wartime Prime
Minister Winston Churchill, whose
leadership during Word War II belied
some of the stereotypes about 'fat
people'. Indeed, his appearance only
increased his 'bulldog' credentials.

considered to be a 'bad' thing. They lack 'moral fibre'. They need to gain self-control. All these terms suggest that 'ugly as sin' is not an outdated attitude. As with all the previous research, this assumption is not necessarily true. If we consider a photograph of Sir Winston Churchill, Britain's famed and widely admired wartime Prime Minister, we see a pudgy face, weak chin, and small eyes hidden by folds of fat. We see a man who, in the prejudice of vanity, would be considered to be morally repugnant. Yet he is widely regarded as having helped to save the western world from Nazi imperialism. A person's moral character cannot be read from their appearance, even though we sometimes assume it can be.

THE TYRANNY OF BEING 'NORMAL'

Then there are people who are 'out of the ordinary' in a more extreme way. And although their differentness is manifestly not 'their fault' in the way that obesity can be considered to be, they are still sometimes blamed and persecuted for their condition.

Vicky Lucas has a rare condition called Cherubism – a medical disorder, which means that from being a normal-looking young child, she grew into an adult with protruding eyes and an oversized chin. She explains that her mother first noticed changes taking place when she was about four years old. 'She was brushing my teeth and noticed that my teeth were kind of higgledy piggledy around my mouth, you know, not quite normal.'

At first the differences from other children were slight. Vicky was quite happy at primary school, and had friends. But research has shown that children with disfigurements of the face attract the strongest rejection and bullying from their peers, and by the age of twelve, Vicky's condition had become worse. 'My jaw started to get bigger, and my eyes started to protrude as well. Then a lot of my friends started becoming quite interested in boys, and fashion, and I found a lot of them weren't happy to be seen with me any more.'

Studies have confirmed just how important it is to adolescents to be associated with attractive peers, and this teenage desperation about being seen with the 'right' people provoked a hostile reaction – Vicky found that one of her friends turned on her. 'We were very good friends up until the age of thirteen, and then in secondary school she actually became one of my worst bullies. I think it was all about being popular. When you're a teenager you want to be popular, you want to be normal, you want to be in, in with the crowd. The last thing you want to be is different, which is what I was, and she didn't like that.'

Some adults still find it difficult to be with individuals who have a facial disfigurement, and studies have shown that they will, for example, avoid sitting next to them in public places. But today Vicky has good friends for whom her facial appearance is unimportant. She also has pockets of time where she will feel she is living the life of a 'normal' 21-year-old college student. Her intelligence, her lively personality and her interests shut out the physical disfigurement. She actually forgets about the way she looks. 'Sometimes I'll be walking around and I'll think that, you know, I'm the hottest chick in town!' she laughs. 'Or I'll think I'm pretty cool! I'm an "OK babe", and then somebody will give me a really bad stare. And then I'll remember that I don't actually look that way.'

She has considered corrective surgery and at this point she has rejected it. Not only would it be risky and painful, with limited chance of major change, but it would also be a way of giving in to people who say she should be 'normal'.

Her condition has led her to consider more closely than most of us what it means to be 'normal'. She has some trenchant views. 'The people who call me names, and shout out at me, "Why don't you go and see a plastic surgeon?" – these are all people who society considers to be normal. So is that what I'm supposed to be like? Should I be like these normal-looking people who call me names? Unfortunately society has prejudices and discriminates against people with disfigurements, but I think it's those attitudes that need to be changed, not necessarily our faces. Why is it that society has such negative feelings towards disfigured people? What is this fear that they have?'

Our fear of being 'different', an outsider to the group, is deep-rooted. For many people the trouble with being a 'normal' person, psychologically as well as physically, is that it requires hiding a lot of oneself. Presenting an acceptable persona to the world all the time means that aspects of our inner selves are not to be acknowledged. These aspects of the self, repressed because they do not fit the persona we want to project, are called by Jung the 'shadow'. They gather strength by not being expressed, and influence our lives from the unconscious. They can poison our psyche and become dangerous.

Vicky appears to be challenging the value that some people place on being 'normal'. By not undergoing plastic surgery in order to 'fit in', she is not playing the game of vanity. Some people seem to assume that she has a duty to undergo plastic surgery so that they will not be offended and upset by her differentness. They are threatened by her apparent willingness to accept herself as she is. She seems liberated from the negative, fear aspects of vanity, and this provokes anger.

The 'normal' people who torment Vicky are revealing dangerously unhealthy shadows.

Above Vicky Lucas, who has a rare condition called cherubism.

Of course, the people who openly berate Vicky are few and far between. As she says, most people are nice. And she lives not as a victim, but as a person committed to asserting the rights of people who do not look 'normal' to not be treated as 'abnormal', as outsiders. She has had to transcend the dark side of vanity and come to value herself as a person. 'I am who I am, and I accept that,' she says.

Today we are becoming more aware of these prejudices. In Britain, the Disability Discrimination Act legislates against discrimination on the basis of facial appearance. Nichola Romsey and Ray Bull, active researchers in this area, have collaborated with the UK charity 'Changing Faces' to raise awareness of the feelings of individuals with a facial disfigurement, and to help others to overcome their prejudices. It is always felt that vanity is self-love. In fact, it seems to be the reverse – it is an attempt to compensate for our lack of confidence, or real acceptance of ourselves. If we were truly comfortable with ourselves we would not need to spend so much time preparing our faces to create a favourable impression on other people. And a danger of excessive vanity is that we finish up being too dependent on other people's opinions of us, making us more fragile and more likely to bolster our feelings of self-worth by discriminating against others.

A more general danger of being overly concerned with managing our appearance, and the impressions we make on others, is the mistake of identifying ourself with our persona. The persona is not the whole person – it is only that part of us that fits a particular role. Identifying too strongly with our persona means that we lose contact with the deeper sources of our own being. Life becomes one continuing series of role-plays, behind which our true self is denied.

Vanity can be a journey away from ourselves, an attempt to be someone we are not. When we look in the mirror tonight, as we go to bed, perhaps we should try to see our face as it is, and not as we automatically revise it in our mind's eye. Does our face reflect our true self? Is the face looking at us, really us?

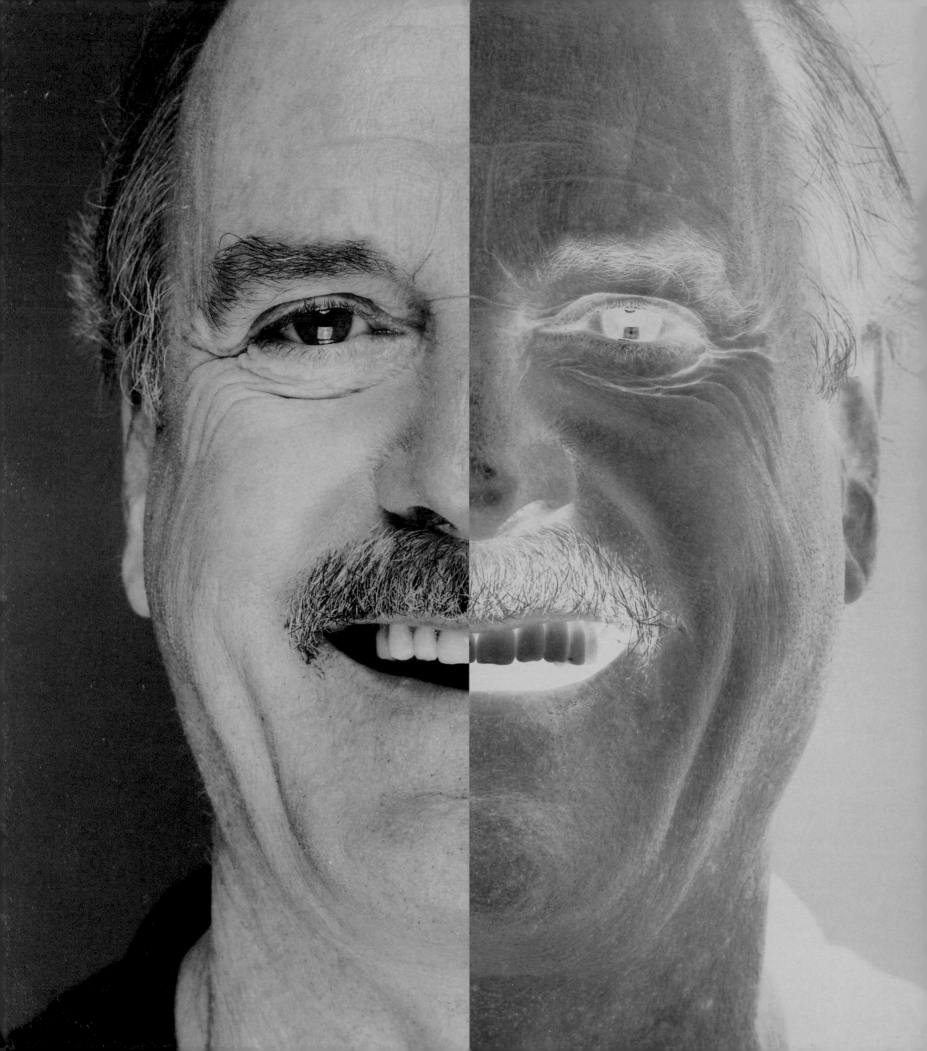

fa|me

CHAPTER 6

Below The poster for the film
Notting Hill.

Opposite Supermodel Christy
Turlington, the face (and body) of a
Calvin Klein advertising campaign.

We recognize people by

their faces. But the faces

of celebrities have a

special importance. They

are purposely glamorized

so that we can project

our hopes, dreams and

fantasies on to them.

RUISING DOWN SUNSET BOULEVARD IN LOS ANGELES IS LIKE HACKING THROUGH A FAME JUNGLE. The faces of film stars sit on colourful movie posters, perched in a tangle of enormous billboards. Hollywood – driven by the engine of cinema – is all about faces: Tom Cruise. Julia Roberts. Leonardo DiCaprio. Madonna. Mel Gibson. Elizabeth Hurley. John Cleese...

Of course, we do not need to go to Hollywood to see these faces – in everyday life we cannot escape them. Magazine racks are loaded with glossy covers clamouring for our attention. The same celebrity faces stare, smile,

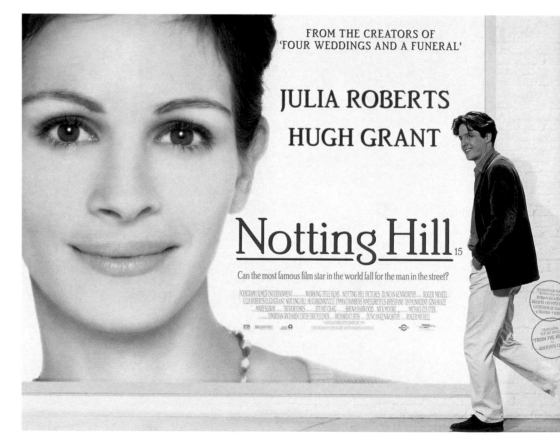

pout and pose. They are available for photo opportunities in newspapers, for chatter in gossip columns, and for lending their features to sales campaigns for perfume, clothes, cars and food, as well as starring in movies. Famous faces everywhere. And they are doing very well financially from being famous, thank you.

The big question is, why? What is it about these people that persuades us to give them such power and influence? Does who becomes famous say something about the values of contemporary culture? There are famous politicians and sports people too – but most of the celebrity faces belong to people in the entertainment industry. Famous dentists and electrical engineers are thin on the ground. So why are entertainers such cultural icons? Has fame been trivialized?

Calvin Klein

Right Eva Herzigova seems very happy with a larger-than-life poster of herself.

Opposite Andy Warhol's silkscreen print of Elvis Presley, a timeless icon.

And why their faces? We know that the face carries more detailed, visible 'information' than other parts of our body – but is there a deeper reason for filling our lives with the familiar features of the famous?

We used to know the answers to these questions, but our understanding of the nature of fame has slipped beneath our awareness. We mistakenly think it to be a contemporary issue only, for fame feeds on the mass media spawned globally in recent decades. But we need to go further back in our cultural

history to explore the roots of such a seemingly shallow phenomenon. The word 'icon' comes from the ancient Greek term for image – and the original meaning is still present in the ikons of the Greek Orthodox Church, portraits that are possessed by the power of spirit and divinity. This sacred aspect of 'famous faces' turns out to be significant, for the story behind fame returns us to the power of the human face as a doorway to the soul.

The story has a number of ingredients – our emotional need for fantasy, the nature of tribal chiefs and shamans, the invention of photography, the advent of the movie close-up, the faces of baddies in animated films, and a Tibetan mask. Let us start on firm ground, with some history.

Kings and queens

Images of our prehistoric ancestors were painted on the walls of caves in Europe. In ancient Egypt, portraits of the mummies were featured on their sarcophagus. Around the world, death masks were carefully preserved. Almost all of these early images were of kings and queens, princes and warriors. It is only in more recent times, as we shall see, that fame has become more 'democratic', with film stars being photographed more often than political leaders.

Today, in an age saturated with the visual media, we are familiar with thousands of famous faces: people we have never met, yet whom we recognize in magazines and on television. But it was not always this way. It is easy to forget how very few faces were famous in the past. Even as recently as the eighteenth century, many rural peasants would have known only those people who lived in the farming community, local village or nearest town. There was no 'mass media'. Images of famous faces were confined to the likes of colourful and infamous highwaymen like Dick Turpin on 'Wanted dead or alive' posters, or political rabble-rousers such as Thomas Paine on revolutionary leaflets.

Opposite A coin depicting French king Louis XVI. His distinctive profile resulted in his identification and capture.

In one case, a king may have regretted the prominence of his 'famous face'. In 1792 King Louis XVI of France fled from the Paris revolt in a bid to escape almost certain death by guillotine. To avoid detection, he dressed as a valet. But he got only as far as the small town of Varennes, where he was captured by the local peasants — a postmaster had recognized him from his distinctive profile on the currency. Louis XVI's widely distributed image of his prominent, curved nose betrayed him. His was the most famous face in the land.

In earlier centuries still, the only 'famous faces' familiar to people in everyday life were those of their monarchs, in profile on their coins.

Early leaders such as Julius Caesar asserted their dominion by stamping their likeness onto the coins of their empire. Later leaders had their pictures printed on banknotes. After coins and banknotes, among the elite of society, came the 'currency' of portraiture. In this time before photography, the images of the famous were painted by artists. They were not mass-produced, of course, and were viewed only by those who had access to the palaces and great houses where such works of art were on show. But, even back then, political spin and public relations were rife.

Elizabeth I ascended to the throne of England in 1558. Just five years later her regime was attempting to control her image. Her ministers drafted a law which complained of 'the daily proliferation of unsatisfactory likenesses of the queen' and their 'errors and deformities'. It suggested that artists should all work from a template of the queen's face, and from which they should deviate as little as possible.

This very early 'press office' edict also ensured that she aged less in her portraits than in real life. As a result, foreign ambassadors and envoys were shocked by the appearance of the monarch in her later years. Her teeth were said to be black and rotten, and her skin ravaged and pitted by white lead paint. Nonetheless, in a time of political turbulence and international belligerence, Elizabeth had to project a strong and stable presence. One way to do that was to ensure that her facial appearance did not alter much. The resulting flattering portraits were the Elizabethan equivalent of today's magazine photoshoots of 'mature' celebrities, where the lens focus is so soft that one struggles to identify the person at all.

In more recent times, politicians have used their facial images to assert their authority over their citizens. The Soviet communist leaders had huge statues cast and erected in every major city. These figures became the focus of hatred when the Communist dreams went wrong. The statues were toppled and smashed in the 1990s when the Russian republics went their separate

Above Communist leaders had their statues erected as icons, and they were publicly dismantled when the Eastern Bloc crumbled.

During the reign of Elizabeth 1, theatre was considered entertainment for the masses as well as being appreciated by wealthy and elite patrons. It was part of popular culture, and its stars were well known. William Shakespeare had a reputation even then as a leading playwright, producer and actor. Yet only one reasonably authenticated portrait survives of him. Times have changed. Today the purveyors of popular culture, people who write, direct and star in their own films, are among the most frequently photographed of famous faces.

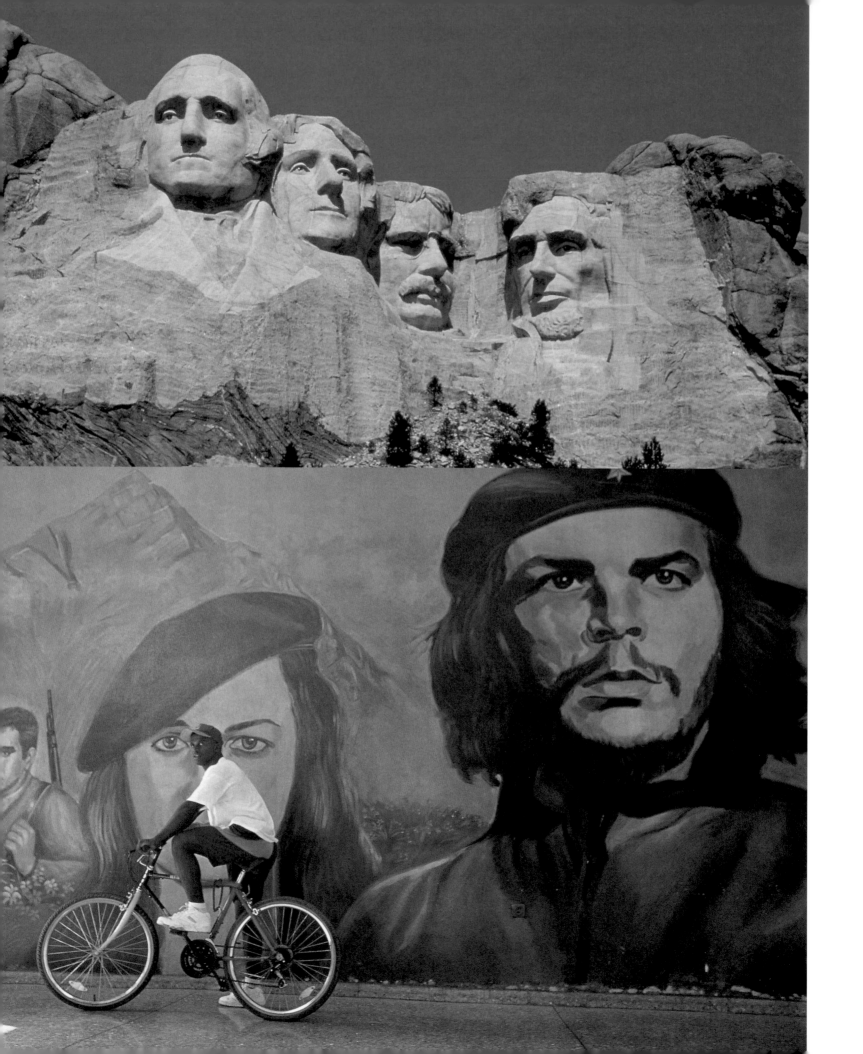

ways. But other arrogant testimonies to power have arisen since in various parts of the world.

Modern western politicians manipulate the presentation of their faces, too. Presidents and prime ministers are the products of a television age. We see their moving images frequently, usually in carefully staged 'photo opportunities' designed to present them in a favourable context.

The importance of projecting a positive image in the fame stakes of politics was confirmed forever in the 1960 contest for the American presidency between John F. Kennedy and Richard Nixon. After losing the election largely because, in his televised debate with Kennedy, Nixon had failed to wear sufficient face make-up to project an attractive presence, he famously admitted, 'In the modern presidency, concern for image must rank with concern for substance.'

So for centuries, people around the world have identified with a manufactured image of their leaders. But at least their fame was understandable. They were leaders either by inheritance, by military power, or because they had been voted into office. They were famous almost by definition. Their role required that they be placed 'on a pedestal' by the masses and be known as a famous face, whether they were loved or loathed, respected or reviled.

The way these leaders manipulated the presentation of their 'famous faces' tells us something about the ubiquitous nature of public relations. However, it makes sense for kings, queens and presidents to be famous. Why film stars? To explore this issue we need, surprisingly, to return to our tribal past.

Tribal imagination

When we lived in tribal societies, we had three kinds of leaders who held special 'star' status in the community. One was the chieftain, the political and ceremonial head of the community. Another was the warrior, the leader into battle. The third was the shaman, the spiritual and healing guru of the tribe. Over the centuries, we have remembered the significance of the chieftain – the famous faces of presidents and prime ministers are on today's magazine

Opposite top Faces of American presidents carved into Mount Rushmore, USA; **bottom** Che Guevara was one of the revolutionary heroes of South America – his face became a potent icon.

covers. We have remembered the warrior, not only as military hero but in the 'ritualized combat' world of sports. Top sportspeople like Michael Jordan and Muhammed Ali fulfil for us the need for heroes in competition for whom we feel loyalty and support. Today the top sports stars are rich and famous.

But we have forgotten about the shaman, whose inspirational role was to journey to the spirit world, and bring back wisdom and blessings to the people of his community. Entranced by his spirit helpers, he enacted for the people the stories of his sacred sojourns to the Otherworld. He took on the spirits' facial expressions, and spoke in their voices. So potent was the shaman's face in embodying the spirit world that in some ritual dramas his face could be represented as a mask – and the spirits entered it of their own accord.

Opposite The ancient, huge carved faces of Easter Island, Chile.

One such ceremony has survived into recent times as a Buddhist ritual. A Tibetan mask is taken out of its shrine once a year and set up overnight in a locked chapel. Two novice monks sit up all night chanting prayers to prevent the spirit of the mask from breaking loose. For miles around, the villagers bar their doors at sunset and no one ventures out. Next day, the mask is lowered over the head of the shamanic dancer who is to incarnate the spirit at the centre of a great festival. The participants in this ceremony are in no doubt that he is presenting a literally awesome face. He is incarnating the face of spirit. These spirit masks are a way of making tangible what we, in modern western society, leave unsaid, sometimes unrecognized, often not understood: the hidden forces of our lives.

Above A ritual dance and the possession of the performer by the spirits on Siberut Island, Indonesia.

The shaman's performances, enacting stories of his encounters with the spirits, provided for the tribe an entertainment with sacred intent. In tribal

Above The mask embodied the spirit-face of the shaman.

Opposite The huge, imposing face depicted on a Native American totem pole, in Canada.

times these living myths were regarded as gifts from the gods which could be understood only from within the imagination. Then, the imagination was for us at least as significant as the pragmatic 'reality' of the everyday.

We no longer understand this view. Today we tend to think of the imagination as an indulgence, as recreation rather than business. But psychologists and psychiatrists are acutely aware of just how important the imagination is to our emotional well-being. For our effective functioning we need both the reality-testing, pragmatic, empirical mind of analysis, and the intuitive faculty of fantasy and visioning. If we had to live our lives locked totally in the rational, without access to fantasy or imagination, we would all go mad.

Kings and queens do not have fans, they have subjects. But the shaman had 'fans'. The word fan derives from the Latin *fanaticus*, meaning 'driven to a frenzy by worship of the divine'. But when the shaman was either outlawed in favour of priests of organized religions, or displaced by the objective world of science and technology, who did we have to take us on journeys of the imagination? The answer is the actor.

Performance without the spiritual element went 'underground' in *commedia del'arte* improvisational theatre, the straight theatre of Shakespeare and his successors, vaudeville and circus. But it was not until modern times that the performer again took on something of that magic of the original shaman. The film and TV stars of today work in a medium that has the power, glamour and impact of the tribal ritual drama. The *'fanaticus'* has been transferred to modern entertainers, whether they are worthy of it or not.

This is no longer a sacred process, of course, but it still carries enormous importance. The imagination is something we need at the deepest levels. But because we have forgotten the nature of this need, and the origins and importance of what shamans used to do, it has left us unable to explain why we pay actors up to $20 million for ten weeks' work on a feature film, and why they are literally 'hero-worshipped', with their faces prominent in all the media.

The simple reason for this is that we need them. Hard though it is to admit, the multi-billion-dollar Hollywood fame machine would not exist unless we were willing accomplices.

Nevertheless, it still seems a big step from sacred spirit journeys with shamans to the 'worship' of soap opera stars. How did this come about?

Right An elaborately made-up
performer in Chinese opera.

Below Terracotta clown mask from
Sardinia, dated *circa* 6th century BC.

The invention of photography

Several technological breakthroughs have changed the nature of fame. The most important was the invention of photography. The earliest daguerreotypes – a method of photography where silver iodide is exposed on a copper plate – were made in 1837. It was to have an enormous impact on the role of the face in popular culture, for it made possible the mass-production of images. Suddenly, famous faces were everywhere. In the mid 1800s there was a raging fashion for *cartes de visite* – small cigarette-card-sized photographs on cardboard – that depicted kings, queens, generals and clergymen. Collecting these became something of a mania. And among the most popular subjects for these picture cards were vaudeville stars!

Photography also democratized fame. Before this, the only people who could afford to have portraits of themselves were the upper echelons of society. But suddenly anyone could have their image reproduced. And if they were potentially famous, then many duplicates of their image could be manufactured to make it ubiquitous.

When Andy Warhol, 1960s pop artist and media supremo, proposed the idea that we would all be famous one day, even if only for fifteen minutes, no one expected it to become literally true. With the Internet now spawning tens of thousands of new personal websites each year, Warhol's prophecy begins to loom large on the horizon. So while photography raised the profile of people who could connect us with our imaginations, it also began the process whereby the notion of fame became diluted due to sheer exposure.

The movie close-up

In theatre, how well we could see the faces of the performers depended entirely on where we were sitting, whether at vaudeville or more serious drama venues. In the 'gods', high up in the balconies, we were too far away to take in much detail of their expressions. In the front stalls we could see a lot more. When movies first started they followed theatrical convention and were shot in one take, from middle-distance. And then a big development happened – the introduction of the close-up. Fame was never to be the same again.

The faces of film actors were blown up to enormous size on the screen. The effect was two-fold. Firstly, it made the actors even more like 'stars' than they were before by projecting their faces on such an awesome scale. It flattered their importance, and made a big impression – literally as well as metaphorically – on the minds of the mass movie-going audience.

Secondly, the huge close-ups were visually the equivalent of gazing at someone's face from a distance of about six inches. The only occasions on which people had previously encountered someone's features at such close range were highly intimate. So cinema close-ups created a close, personal bond between the members of the movie audience, and the famous star they so admired.

The widespread fame and 'fanaticus' which gathered around film actors was a new phenomenon. It was also very profitable for Hollywood. It maximized audiences for films by drawing them to see their favourite stars. And to exploit the financial fortunes to be made from fame, film actors were soon subjected to the same 'spin' that political leaders had employed for centuries. The doyen of famous faces on the silver screen was Max Factor, an escapee from the Russian Imperial Court. He was a make-up magician for thousands of women in the 1920s and 30s, the early years of Hollywood. He created the looks of a generation of screen stars – everyone from Clara Bow to Greta Garbo and Lana Turner. Max dusted gold into Marlene Dietrich's wigs, and conjured the bright platinum-blonde hair dye for Jean Harlow. He also created make-up that looked flawless in close-ups. The make-up was also marketed for the generations of women who grew up wanting to look like the movie idols. The selling slogan was 'The make-up to the stars and you!'

Sometimes the make-up was not itself sufficient to create the glamorous image of the famous faces. From the 1920s onwards the major Hollywood studios employed teams of artists who re-touched photographs of the stars. Nowadays air brushing can be done at the touch of a computer button, but then it involved scratching the emulsion off the photographic negative and drawing in new features, removing wrinkles and blotches by hand. Famously, there are two identical photographs of Bette Davis in existence, one before re-touching and one after. The contrast is striking. In some cases the retouching was done in the flesh rather than on film. Since then, the pampering of famous faces has been relentless, turning

Opposite Film actress Lana Turner, whose career peaked during the 1940s.

Above Film actress Rita Hayworth in an early poster advertisement for Max Factor make-up.

actors into stars who could be admired and sought after by millions of fans.

The process of glamorizing the stars continues today. The *Oxford English Dictionary* confirms the original meaning of 'glamour' to be a magic enchantment or spell – and an early use of the word was 'When devils, wizards or jugglers deceive the sight, they are said to cast glamour o'er the eyes of the spectator'. 'Glamour' takes the performer into the realm of magician. And the use of close-ups continues unabated.

David Stewart, one of the producers of *The Human Face* TV series, compared close-ups in two films. The 1939 *Wuthering Heights*, starring Laurence Olivier, had 22 per cent of its shots in close-up. The figure for the 1998 film *Shakespeare in Love*, starring Gwyneth Paltrow, was 51 per cent. This more than doubling of close-up scenes may be because they are found to be so effective in locking us on to the face of the film's stars, and engaging us in the story. Or it may be that we are so used to seeing the faces of star actors blown up to enormous size that the technique needs to be used on us more intensely, in order to evoke the same reaction as before.

Also, in more recent times, television has become a prime medium for maximum distribution of the faces of the famous. Television uses close-up shots a lot. The images are on a small screen but, viewed in the intimacy of our living rooms, they have been 'allowed' into our personal space, and therefore we form emotional connections with them. As a result, television performers from popular, long-running shows such as *Friends* have become recognized around the world.

Positive Projection and heroes

We might feel intimately bonded with movie stars because of the close-ups and publicity, but why are we are so gripped by their screen characters? What is it, beyond merely enjoying the film entertainment, that leads us to make the actors portraying those characters so enormously famous? It is a combination of our unconscious need for heroes, and the actors' ability to bond us emotionally to the characters they portray.

There is a process that psychologists call Identification, which helps us to understand this phenomenon. When we watch a movie, and connect with the

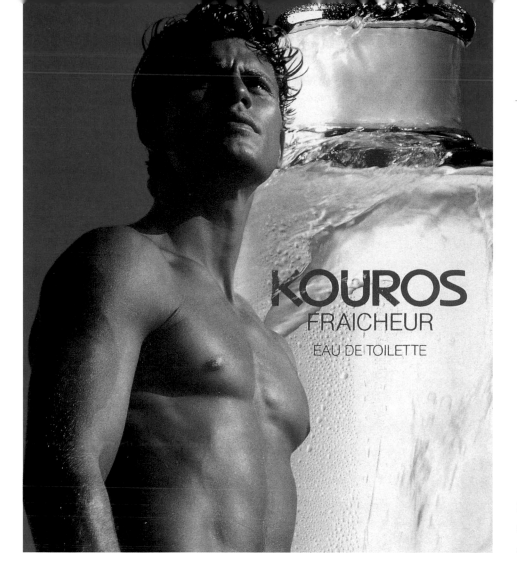

Left This aftershave advertisement associates the product with strength and rugged good looks.

face of the actor in close-up, we have a particular and personal response to the film character. We might imagine ourselves to have attributes of that role. This 'identification' is exactly the reaction sought from us by the makers of advertising campaigns.

Because the faces that front commercials need to connect with us very quickly – in fifteen seconds of television watching, or the time it takes to glance at a magazine page – casting agents rely on our identifying with stereotypes: faces that almost always evoke a particular reaction. Casting directors know precisely which actors and models will sell which product. The faces that best fit the brief depend on how the commercial director intends that we should relate with the characters in the advertisement: admiration, pity, lust, humour, fondness and so on. Research has shown, for example, that mature rather than young-looking faces are more persuasive when selling a product based on their expertise.

And a commercial for a DIY product, for instance, needs the viewer to identify with the character in a comfortable, unchallenged way. The face is

not meant to be exciting! He might look like a 'Dad', but not too interesting. Good-looking, but not too handsome. No distractions like a beard, or unusual features. Frequent smiles.

Some people become 'famous faces' as a result of work in commercials. But usually it requires more sustained performances, on television or in films, to attract the level of identification that secures 'fans'. This takes place through a stronger version of identification, which psychologists call Projection.

In Positive Projection we go beyond simply imagining that we might be like the character. Rather, we attribute to the character qualities that we would like to have. We may overdo it, and exaggerate the positive qualities of the character. But projection often has an element of hero-worship.

We connect deeply only with those characters onto whom we can project personally, and who for us represent a kind of hero. We are projecting, on to the image of the character, a large part of ourselves, and this bonds us with a character – and the actor playing that role – powerfully.

To activate the process we need what is called a 'hook' – an element or feature of the character onto which we can hang our psychological hopes and dreams, fantasies and desires. Film producers, screenwriters and directors are skilled at providing us with the kinds of stories that 'hook' us. But beyond the story of the film, it is the faces of famous actors that provide ideal hooks for Positive Projection – especially given that they are usually attractive, or even

Left Fava beans and a nice Chianti, anyone? Anthony Hopkins created a chilling face of evil – Hannibal Lecter in *The Silence of the Lambs*.

beautiful, and we see them in huge close-up. So we project parts of ourselves onto the character, and this then becomes transferred to the star actor who portrayed that role.

The faces of celebrities in magazines become a kind of fantasy world of 'hooks' onto which we can Project. In this way a celebrity is an artificial construction. The public persona of 'stars' like Jude Law, Madonna, Ralph Fiennes, Gwyneth Paltrow and Will Smith may bear little relation to their private personality. The star actor is a real person, but perceived through the prism of characters they play, and the studios' publicity machines. The star that the public sees on film and in the magazines is a *version* of a real person, presented and perceived in a particular way to fulfil the wishes, desires and fantasies of fans.

Positive Projection onto famous faces often makes the stars the object of fantasies. We all have an inner stream of consciousness that mixes together images from our personal past, anticipated futures and the realities of the everyday world. We live simultaneously in the world of fantasy and reality. Daydreams are a kind of intermittent internal reverie. They are full of 'unrealistic' images that are not directly connected with the 'objective' world in which we live. In this private, inner world, we daydream about love, success, sex, happiness, material wealth and revenge.

For many people the famous faces of 'stars' and 'celebrities' feature in these fantasies. And asked to list people one either knows or has heard of,

Below Mass murderer Ted Bundy, who
looks disappointingly 'normal'.

Opposite Jack Nicholson, an actor
famous for his portrayal of evil
characters, in a still from the film
The Shining.

many people can list as many film and television characters as individuals in
their 'real' social world. Their faces leap to mind, and stars can take on an
importance almost as strong as someone we know.

Part of the price celebrities pay for being a famous face is an insatiable
curiosity about their private life. The general public is fascinated by, and
confused by, the relation between the 'real' person and 'star'. It is the star to
whom they are attached. But they think they 'know' the real person.

Sometimes the stars are themselves confused by the public's reaction to
them. They know that the public do not really know them as people, and yet
sometimes act as if they do. And yet their fans also seem to have expectations
that they will behave like the characters they have played. This can be
uncomfortable for the actor, especially where he feels himself to be clearly
different from the character.

For actors, continuing fame depends on being predictable. They can
show off different facets of the persona they have come to represent, or even
try, as some do, to present their 'real self' to the public. But they cannot easily
abandon the persona altogether. When fans think of actors such as Christian
Slater, Jennifer Lopez, Brad Pitt and Jennifer Aniston, it is the 'persona', or
public face presented in performance, to which the public have become
attached.

Negative Projection and villains

If we Positively Project onto attractive heroes, what then of villains? In real
life the general public is endlessly fascinated by famous villains. But often the
ingredient we are looking for is missing. With the notorious exception of
Charles Manson, the famous faces of even mass murderers are often
disappointingly 'normal': they do not look villainous enough. The press tries
to exploit our obsessions with huge headlines above the bland features of mass
murderers like Ted Bundy: THE FACE OF EVIL.

We may abhor what these people have done, but looking at their image
does not carry the weight of their actions. They do not have staring eyes and
a cruel mouth. Sometimes the press try to feed our expectation that baddies
will look a certain way.

The reason we like 'baddies' to look bad is because of a psychological process called Negative Projection. This works differently from the positive variety. It is a process in which we attribute to others aspects of ourselves that we would like to disown. It relieves us, albeit temporarily, of the knowledge that we are harbouring undesirable traits, because we convince ourselves that it is other people who have them.

Left to right Evil faces: Captain Hook from the animated Disney film *Peter Pan*; the Wicked Witch of the West from *The Wizard of Oz*; Robert De Niro in *Cape Fear*; and television's most famous bad guy, J.R. Ewing (Larry Hagman) in *Dallas*.

This is a largely unconscious process – we are not aware that we are doing it. But for it to work well, the people on to whom we want to Negatively Project need to have the right 'hooks'. They need to be effective icons for us to use them to dump our negative aspects.

While real villains are often disappointing in this regard, we can rely on films to provide us with really good hooks for Negative Projection. The chilling face of Anthony Hopkins as Hannibal Lecter in *The Silence of the Lambs*, for example, was perfectly hateful – and very successful. And he created a visage specifically for the role, knowing that the film required that we detest the character.

As we saw in Chapter 4, animated characters can work on our emotions every bit as well as flesh and blood versions – witness the heart-tugging Bambi. Studios such as Dreamworks and Disney use artistic cunning and technical wizardry to make sure that all their characters' faces speak volumes to people

all over the world. And their villains have particular looks, rather like stereotypes, which function as perfect hooks for Negative Projection. They let you dump all your negative stuff without quite realizing you are doing it.

These cartoon baddies tend to have big noses, uneven features and pointy eyebrows – all features that deviate from the ideal 'beauty' we associate with good behaviour. Captain Hook, Cruella De Vil and Snow White's stepmother all have pointy faces. There is surprising uniformity in the faces that are considered 'evil' all around the world.

Above Sylvester Stallone's action films, such as *Rambo*, attracted huge audiences with simple characters and storylines of mythological resonance.

We saw in the last chapter how attempts to correlate facial appearance with morally good or bad behaviour have been discredited scientifically. But the point is, such features enable us to project our negatives on to that character. Although it is not a true connection, our mind prefers predictable images on to which to project.

Cesare Lombroso, an Italian doctor in the nineteenth century, tried to define the physical characteristics of evil. He examined the faces and skulls of convicted felons and claimed that the signs in the face that indicated villainy included enormous jaws, high cheekbones, extreme size of the eye orbits, handle-shaped ears, extremely acute sight and tattooing. This description of the archetypal villain re-surfaces in Bram Stoker's *Dracula*. The aquiline nose and pointed, elongated ears of the Count are taken almost word for word from Lombroso's account.

Tribal dreams

It is easy to dismiss popular entertainment as frivolous, even decadent. But when in the 1970s films such as Sylvester Stallone's *Rambo* and *Rocky* exploded onto our screens and filled cinemas around the world – to the dismay of many because of the violence depicted in them – it underlined how little we understood about what these dramas represent for us *psychologically*. We no longer live as small, homogeneous communities sharing a way of life and common values, traditions and stories. But many of our emotional instincts are still tribal. We identify, in our dreams and deep unconscious, with heroic storylines, timeless epic struggles and issues of spirit – even though we prefer to give them psychological labels such as 'redemption'.

Film producers and writers around the world consciously strive to mine this deep emotional seam. They construct characters and storylines for films in contemporary settings, but which evoke in us that deep, mythological response. Interestingly, an influential figure in the film industry is Christopher Vogler, who serves as a story consultant on many big films. He wrote a memorandum for studio heads that linked successful feature film characters and plotlines to Greek myths. Studio heads were convinced by his argument that timeless mythological stories underpin many of the most popular films of today. Now studios employ him to look for storylines that are like the tales told by shamans, and characters that conform to the ancient and tribal models of the mentor, hero, threshold guardian, messenger, shapeshifter, shadow and trickster.

Below Ritual wooden mask, used in the sacred dance of Bali.

Rambo, and the action film characters that followed, are like the stock, masked characters in tribal, ritual drama. In a way, these films deal with the deepest psychological realms that used to be the province of the shaman.

Some famous actors play a wide variety of roles. But fame, especially in Hollywood, is often associated with 'personality' actors, who invariably play a character recognizably similar to all the characters they have played before. The earlier years of Hollywood saw personality actors become enormously popular: James Cagney, Gary

Cooper, Humphrey Bogart, Marilyn Monroe, Mae West. In recent decades the likes of Clint Eastwood and Sylvester Stallone usually embody a particular screen character, regardless of their private personality. The roles may change, but the audience always knows what sort of character to expect.

In the film *Braveheart*, Mel Gibson plays a heroic warrior summoning up his Scottish countrymen to defy the invading English. It is not a 'sacred' story, but it has the hero battling on behalf of his community and for freedom, very like the ancient shamanic dramas of struggles against the evil spirits. The characters in this film are not drawn in a subtle and sophisticated way. They are the archetypal faces of heroes and villains, like those in tribal mask dramas. The film takes people on a timeless journey in their imagination, and was a smash hit at the box office.

We used to talk of spirits. We now see these matters in secular, psychological terms. But we still have the same inner selves, relations with others, fears, dreams, passions, personal dramas and issues of life and death as we had in tribal times. It is not hard to realize how the star actor-in-performance is still seen, by fans, as incarnating a 'deity' or 'spirit'. Perhaps we do still need spirits of a kind, manifestations of the forces in and around us, and that is why we make some people famous.

Of course, while the famous faces of film stars represent something deep within us, this is not to say that the films in which they feature necessarily have sacred meaning for us – or even deep psychological insight. And, of course, by making fame such a commonplace, we have trivialized it. The insatiable hunger in the media for famous faces, and the profits to be made from them, mean that we are drowning with these images. We are asked to Positively Project not only onto famous actors incarnating 'archetypal spirit forces', but also onto celebrities who give racing tips, weather forecasts, cooking recipes, DIY instruction and so on. The conjuring of endless celebrity is one of the reasons we have forgotten its deeper significance in our lives. Celebrity is a diluted and even polluted pool.

However, our endless fascination with fame reveals the hold it retains over us. It still survives, hiding its power under the magazine cover and billboard images of famous faces.

Opposite Humphrey Bogart, one of Hollywood's 'personality' actors, usually portrayed the same kind of character in his films.

225

conclusion

ORIGINS, IDENTITY, EXPRESSIONS, BEAUTY, VANITY AND FAME
— WE HAVE EXPLORED THE STORY OF THE FACE FROM THE
PRIMEVAL SEAS OF LONG AGO TO THE WORLDWIDE MEDIA
WEB OF TODAY. What have these explorations shown us about our face and
its significance?

The human face is a dynamic structure, changing over evolutionary time
and in our own lifetime. Whereas the course of evolution is something
beyond our immediate knowing, our experience of our own face is first hand.
Our face is not only a familiar sight to us, it is physically ours alone, and
central to our sense of identity. It defines us to the world, and to ourselves.
This individual identity is crucial to our lives as social animals. It locates us
in our complex social world, and allows us to establish a central base from
which to operate. We have a sense of living 'behind' our face. Our interior,
private selves peer out at the world through the two convenient holes in the
front, our eyes the windows from our inner selves to everything else outside.
And for others, our eyes are the windows to our souls.

Identifying with our face means we tend to take it for granted. When we
look in the mirror, many of us do not see a clear image. Our brains construct,
through a specialized process, the familiar montage of our features. But
familiarity dulls our perception, and we fail to register much of the interesting

detail. We see what we are used to wanting to see. To realize the true shape and structure of our faces, we need to step beyond the automatic defences of our 'vanity system', and experience our faces afresh. In a different way. Our faces are a perpetual mystery, even in the contours of expression, and discoveries about our features are like perpetual enquiries into our identity.

Even if we are sitting alone, an expression may flit across our features – we show our basic emotions in our faces even if there is no one there to communicate them to. But much of our attention to expressions is to do with controlling our own and, especially, reading those of others. We all hope to be sophisticated and sensitive readers of expressions. However, we are not as effective at it as we would wish, and we tend to make lots of mistakes when interpreting others' faces. Even people whose work involves judging others by their expressions – secret service agents, judges, police officers – find it difficult. But through practice they become better than us, and can read faces a little more accurately. Observation is the key. Most of us do not observe faces very closely. But we can train ourselves.

We do not pay attention to most of the faces we see, because we walk around half-blindly, attending instead to our stream of inner thoughts. But it can be a rewarding exercise to free our mind occasionally to concentrate on looking at others in public. If you are discreet, people can easily be observed without offending them.

About half an hour in the streets of a town or in a shopping centre is usually ample time for finding faces that stick in the mind. Their faces begin to clarify and individualize. The richness of their variety is striking. So too can be the extent to which the faces give you a feeling for the person. They are like life crystals, an essence of a person that can be developed like a hologram into a full-colour personality portrait in our mind's eye. The way people habitually hold their faces seems almost like conscious expressions, which become more 'set' with age. And even in very young people, the face is already reflecting the deep pools of experience within. The personal mannerisms of the face can be vivid give-aways of personalities, transparent beneath the flimsy masks of the public persona. We can never completely hide who we are, or what we are. Our face reveals what our life has done to us, and the kind of person it has made us.

Looking helps us to see these processes of revealing and hiding. We come to spot more readily an insincere performance. Most of all, we come to understand others a little more.

In everyday life, people who are beautiful get looked at more often. And we have seen that, all over the world, people seem to agree on which faces are considered beautiful. But beauty is a matter not of simple aesthetics, but also of the heart. The character of a person expressed through a lively and mobile face is what really makes them attractive. The way we see people's faces is an active process. We are not passive recorders of the world around us. We select when we look – necessarily, because that is the way the brain works. In seeing, the eyes supply the brain with sensory data about the external world. But to make any sense of this data, the brain has to interpret and organize it. We may be able to agree on judgements of a 'physically beautiful face' but, beyond casual romance, our more committed choice of lover, partner or spouse is affected by our mind's filter of beliefs, prejudices, passions, fears, interests and blind spots. We see what our needs and desires allow us to see.

In other words, in everyday life, we find attractive the people we like. On looks alone we might not consider ourselves to be contenders for the magazine cover treatment, but almost certainly someone appreciates our special beauty.

While personal attraction has a deep side, it also has the glittering surface so beloved of the cosmetics industry. Vanity has probably shaped our haircut.

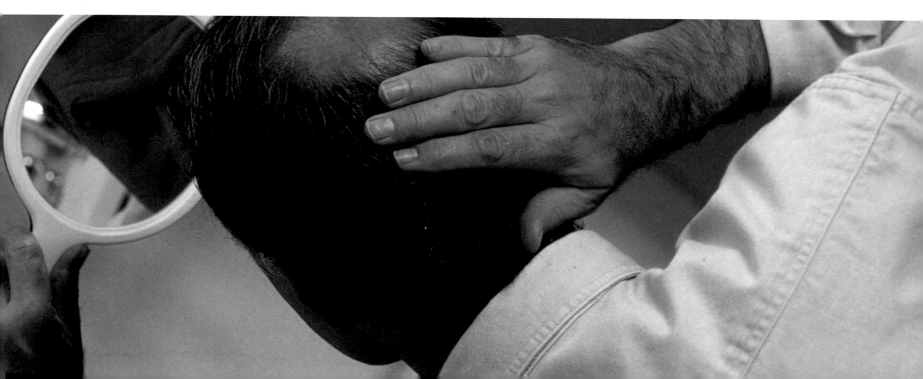

At least! Attending to our personal appearance can be a matter of healthy self-expression. But it also seems to be about 'belonging', marking our allegiance to one social group or another. An obsessive self-regard and concern with appearance preys on our weaknesses. To make ourselves feel better, we put others down.

Vanity can be harmless, as self-expression. And it can be dangerous, as a prop to our need to belong. We need to be aware of our own needs and insecurities and how they affect our own sense of insiders and outsiders, to ameliorate the unconscious prejudices that can arise from them.

Our faces, pampered or not, feature in the memory banks of hundreds of people. And alongside our remembered image, people's minds are crowded with a multitude of faces of people they have never met, but could recognize – celebrities who are 'famous faces' in the media. The face is such an important item on the human agenda that we lock on to those that have special significance for us. Politicians perhaps, sports stars certainly. But actors especially. They are people who take us on journeys into our own imagination, and provide famous features onto which we can project our inner desires, hopes and dreams. This is an ancient process, stemming from tribal times. But we have forgotten the original importance of such catalysts of the imagination. Famous faces are a dime a dozen today. But knowing why some are important for us tells us a lot about ourselves. What qualities do we admire and aspire to, what do we despise and refuse to recognize in ourselves?

Judging people by their faces is necessary when we need to recognize who they are, or understand their expressions. Judging people's facial attractiveness is a combination of biological and personal factors – and can lead us astray when we imagine we can know others by their beauty, or lack of it. The face is front and centre in most of the things we do.

Within our life span we meet others face to face, we face up to problems, save face, face the future. Our face points us forward into life, intrinsic to our identity both personal and social, moment to moment, from birth to death. This story of the human face has sought to refresh and inform our familiar view of our own face. For as we understand more about it, we thereby understand more about ourselves.

bibliography

INTRODUCTION

The face on the Makapansgat cobble: R. Bednarik (1999). 'Der Kiesel von Makapansgat. Fruheste Urkunst der Welt?' *Anthropos* 94: 1/3 (with English summary), and English version 'Makapansgat Cobble Analyzed' on the University of Melbourne website at www.geom.unimelb.edu.au/aura/MAKAPANSGAT. htm.

American and British studies show that thirty minutes after being born we prefer to gaze at faces more than any other object: C.C. Goren, M. Sarty and P.Y.K. Wu (1976). 'Visual following and pattern discrimination of face-like stimuli by new born infants.' *Pediatrics* 56: 544–9.
M.H. Johnson, S. Dziurawiec, H. Ellis and J. Morton (1991). 'Newborns' preferential tracking of face-like stimuli and its subsequent decline,' *Cognition* 40: 1–19.
G.E. Walton, N.J.A. Bower and T.G.R. Bower (1992). 'Recognition of familiar faces by newborns.' *Infant Behavior and Development* 15: 265–9.

We come into the world primed to connect with faces: J. Morton and M.H. Johnson (1991). 'CONSPEC and CONLEARN: A two-process theory of infant face recognition.' *Psychological Review* 98: 164–81.

Our brain recognizes a montage of features as a face: A.W. Young, D. Hellawell, and D.C. Hay (1987). 'Configurational information in face perception.' *Perception* 16: 747–59.

CHAPTER ONE: ORIGINS

Evolution of the face from reptiles to mammals, and to apes: J. Cole (1998). *About Face*. Cambridge, Massachusetts: MIT Press.

'Lucy', the most famous of our primate ancestors: D. Johanson and B.Edgar (1996). *From Lucy to Language*. London: Weidenfeld & Nicolson.

Development of diverse appearance of early humans: R. Lewontin (1995). *Human Diversity*. New York: Scientific American Library.

How does a baby's face develop?: M.A. England (1996). *Life Before Birth: Normal Fetal Development*. YearBook Medical Publishers.

Skin, and melanin acting as a natural sunscreen: A. Damon, ed. (1975). *Physiological Anthropology*, pp. 51–5. New York: Oxford University Press.

As humans migrated from Africa 100,000 years ago they encountered colder environments: D. McNeill (2000). *The Face*, pp. 96–7. London: Penguin Books.

The skin of ancient Scandinavians seems to have been too dark to let sufficient sunlight through: J. Kingdon (1993). *Self Made Man*, pp. 234–5. New York: Simon and Schuster.

Our eyelid is composed of skin only one millimetre thick: D. McNeill (2000). *The Face*, p. 27.

London: Penguin Books.

The shape of our noses seems to have evolved in response to environmental factors: D.H. Enlow (1982). *Handbook of Facial Growth*. Philadelphia: W.B. Saunders.

Different nose shapes were attributed by nineteenth-century travellers to fashion: N. Etcoff (2000). *Survival of the Prettiest*, p. 141. London: Abacus.

Lips, teeth, tongue and chin: D. McNeill (2000). *The Face*. London: Penguin Books.

Our faces have evolved also through aesthetic preferences affecting our selection of sexual partners: J. Diamond (1992). *The Third Chimpanzee: The Evolution and Future of the Human Animal*. New York: HarperCollins.

Our diverse appearance around the world is made all the richer by the increasing degree of racial mixing: M. Kohn (1995). *The Race Gallery*. London: Jonathan Cape.

CHAPTER TWO: IDENTITY

Genetic inheritance and facial features: J. Gribbin and M. Gribbin (1995). *Being Human*. London: Phoenix Press.

Why are male and female faces different: L.A. Zebrowitz (1997). *Reading Faces*, pp. 17–18. Boulder, Colorado: Westview/Harper Collins.

Women have lighter skin tone than men: P.L. van den Berghe and P. Frost (1986). 'Skin color preference, sexual dimorphism and sexual selection: A case of gene culture co-evolution?' *Ethnic and Racial Studies* 9: 87–113.

As women get older, their faces become less feminized: L.A. Zebrowitz (1997). *Reading Faces*. Boulder, Colorado: Westview/Harper Collins.

Animals seeing their reflection in a mirror: G.G. Gallup (1982). 'Self-Awareness and the Emergence of Mind in Primates.' *American Journal of Primatology* 2: 237–48.

Human infants seeing their reflection in a mirror: Anthony Smith (1998). *The Human Body*, pp. 79–80. London: BBC Worldwide.

Polaroid photographs of tribespeople in New Guinea: D. McNeill (2000). *The Face*, p. 109. London: Penguin Books.

Prosopagnosia is a disorder in which people cannot see faces: R. Bruyer, ed. (1986). *The Neuropsychology of Face Perception and Facial Expression*. Hillsdale, New Jersey: Erlbaum.

Prosopagnosia occurs as the result of adult brain damage: N.L. Etcoff, R. Freeman and K.R. Cave (1991). 'Can we lose memories of faces? Content specificity and awareness in a prosopagnosic.' *Journal of Cognitive Neuroscience* 3: 25–41.

Identikit and photofit techniques: H.D. Ellis, J. Shepherd and G. Davies (1975). 'An investigation of

the use of the photo-fit technique for recalling faces.' *British Journal of Psychology* 66: 29–37.

To recognize faces we need light and shade, not just outline: D. McNeill (2000). *The Face*, pp. 82–3. London: Penguin Books.

Exaggerated caricatures of faces trigger recognition: S. Stevenage (1997). 'Face facts: theories and findings.' *The Psychologist* April 1997, pp. 163–8.

Processes of face recognition: V. Bruce and A.W. Young (1998). *In The Eye of the Beholder: The Science of Face Perception*. Oxford: Oxford University Press.
K.R. Laughery, J.F. Alexander and A.B. Lane (1971). 'Effects of target exposure time, target position, pose position, and type of photograph.' *Journal of Applied Psychology* 55: 477–83.

We are poor at recognizing faces of an ethnic type not our own: T. Anthony, C. Copper and B. Mullen (1992). 'Cross-racial facial identification: A social cognitive integration.' *Personality and Social Psychology Bulletin* 18: 296–301.
P. Chiroro and T. Valentine (1995). 'An investigation of the constant hypothesis of the own-race bias in face recognition.' *Quarterly Journal of Experimental Psychology* 48a: 879–94.

Machines are quicker at remembering faces than we are: D. McNeill (2000). *The Face*, pp. 83–4. London: Penguin Books.

We can recognize the faces of people we know even under difficult circumstances: H.P. Bahrick, P.O. Bahrick and R.P. Wittlinger (1975). 'Fifty years of memory for names and faces: A cross-sectional approach.' *Journal of Experimental Psychology* (General) 104: 54–75.

We recognize our mother's face very early: I.W.R. Bushnell, F. Sai and J.T. Mullin (1989). 'Neonatal recognition of the mother's face.' *British Journal of Developmental Psychology* 7: 3–15.

CHAPTER THREE: EXPRESSIONS

John Gottman's 'Love Lab': E. Nussbaum (2000). 'Inside the Love Lab.' *Lingua Franca* 10 (2): 1–14.

Children good at recognizing others' expressions are popular: A. Manstead and R. Edwards (1992). 'Communicative aspects of children's emotional competence.' In K.T. Strongman ed., *International Review of Studies on Emotion*. New York: Wiley.

Facial expressions and emotions: J.J. Gross and D. Keltner, eds. (1999). *Cognition and Emotion*. Psychology Press.

The muscles and nerves of the face: J. Cole (1998). *About Face*. Cambridge, Massachusetts: MIT Press.

Effects of alcohol on emotional attention: R. Carter (1999). *Mapping the Mind*. London: Orion.

Mammals evolving more mobile faces than reptiles: J. Cole (1998). *About Face*. Cambridge,

Massachusetts: MIT Press.

Darwin concluded that the basic expressions represented the same recognizable emotions all over the world: C. Darwin (1998). *The Expression of the Emotions in Man and Animals: The Definitive Edition.* With introduction, afterword and commentaries by P. Ekman. New York: Oxford University Press.

Ekman discovered that expressions were a universal language: P. Ekman, W.V. Friesen and P. Ellsworth (1982). 'What are the similarities and differences in facial behavior across cultures?' In P. Ekman ed., *Emotion in the Human Face* (2nd edition). Cambridge: Cambridge University Press.

The seven basic expressions: P. Ekman, W.V. Friesen and P. Ellsworth (1982). 'Does the face provide accurate information?' In P. Ekman ed., *Emotion in the Human Face* (2nd edition), pp. 56–97. Cambridge: Cambridge University Press.

Even infants just 3 months old can recognize basic emotions: G.M. Schwartz, C.E. Izard and S.E. Ansul (1985). 'The 5-month-old's ability to discriminate facial expressions of emotion.' *Infant Behavior and Development* 8: 65–77.

Babies start smiling as a reflex regardless of anything their mothers may do: A. Smith (1998). *The Human Body*, p. 67. London: BBC Worldwide.

Newborn babies look at faces for longer than any other object put in front of them: I.W.R. Bushnell, F. Sai and J.T. Mullin (1989). 'Neonatal recognition of the mother's face.' *British Journal of Developmental Psychology* 7: 3–15.

Newborn infants show a preference for attractive human faces: A. Slater, C. Von der Schulenburg, E. Brown, M. Badenoch, G. Butterworth, S. Parsons and C. Samuels (1998). 'Newborn infants prefer attractive faces.' *Infant Behavior and Development* 21 (2): 345–54.

In the first days of life, the baby will start to mimic the lip and cheek movements of its parents: A.N. Meltzoff and K.M. Moore (1977). 'Imitation of facial and manual gestures by human neonates.' *Science* 198: 75–8.

We make our facial movements slower and more exaggerated to babies than we would towards adults: D. Stern (1977). *The First Relationship: Infant and Mother.* Cambridge, Massachusetts: Harvard University Press.

Communicating exaggerated expressions with babies seems to be practised all over the world: I. Eibl-Eibelsfeldt (1989). *Human Ethology.* New York: Aldine de Gruyter.

A mother will often sustain eye contact with her baby for thirty seconds or more: J.C. Peery and D. Stern (1976). 'Gaze duration frequency distributions during mother–infant interactions.' *Journal of Genetic Psychology* 129: 45–55.

Scientists estimate that there are eighteen varieties of smile: P. Ekman and W.V. Friesen (1984).

Unmasking the Face. Palo Alto: Consulting Psychologists Press.

The Duchenne smile: G.B. Duchenne (1990). *The Mechanism of Human Facial Expression.* Translated by R.A. Cuthbertson. New York: Cambridge University Press.

False expressions are slightly asymmetrical: P. Ekman, W.V. Friesen and M. O'Sullivan (1988). 'Smiles when lying.' *Journal of Personality and Social Psychology* 54: 414–20.

Judging from expressions when someone is telling a lie: P. Ekman and M. O'Sullivan (1991). 'Who can catch a liar?' *American Psychologist* 46: 913–20.

Each culture has developed its own set of rules about how much our face is allowed to give away, or conceal, our emotions: P. Ekman (1991). *Telling Lies.* New York: Norton.

The Japanese make less eye contact in business: D. McNeill (2000). *The Face*, p. 247. London: Penguin Books.

The meanings of head nods and shakes: Desmond Morris (1994). *The Human Animal*, pp. 20–3. London: BBC Books.

The human eye shows more white than that of other animals: V. Bruce and A. Young (1998). *In the Eye of the Beholder*, p. 10. Oxford: Oxford University Press.

Intimate eye contact can be an act of flirting: J. Brophy (1962). *The Human Face Reconsidered*, pp. 22–3. London: Harrap.

How we feel affects the way we blink: P. Ekman (1991). *Telling Lies*, pp. 142–3. New York: Norton.

Interaction via video link is different from face-to-face communication: V. Bruce and A. Young (1998). *In the Eye of the Beholder.* Oxford: Oxford University Press.

CHAPTER FOUR: BEAUTY

All over the world, people seem to agree on which faces are considered beautiful: M.R. Cunningham, A.R. Roberts, A.P. Barbee, P.B. Druen and C.H. Wu (1995). 'Their ideas of beauty are, on the whole, the same as ours: Consistency and variability in the cross-cultural perception of female physical attractiveness.' *Journal of Personality and Social Psychology* 68: 261–79.
L.A. Zebrowitz, J.M. Montepare and H.K. Lee (1993). 'They don't all look alike: Individualated impressions of other racial groups.' *Journal of Personality and Social Psychology* 65: 85–101.
I.H. Bernstein, T.D. Lin and P. McClellan (1982). 'Cross-vs. within-racial judgements of attractiveness.' *Perception and Psychophysics* 32: 495–503.

Two relatively isolated tribes in South America, the Hiwi and Ache: D. Jones and K. Hill (1993). 'Criteria of facial attractiveness in five populations.' *Human Nature* 4: 271–95.

Appearance sensitivity and eating disorders in adolescent girls: C. Davis, G. Claridge and J. Fox (2000). 'Not just a pretty face: Physical attractiveness and perfectionism in the risk for eating disorders.' *International Journal of Eating Disorders* 27 (1): 67–73.

The golden mean: J. Liggett (1974). *The Human Face*, p. 140. London: Constable.
L.A. Zebrowitz (1997). *Reading Faces*, pp. 123–4. Westview Press/HarperCollins.

Ancient formulae do not work on contemporary models: L.G. Farkas, I.R. Munro, and J.C. Kolar (1987). 'The validity of Neoclassical facial proportion canons.' In L.G. Farkas and I.R. Munro eds., *Anthropometric Facial Proportions in Medicine*, pp. 57–66. Springfield: Charles C. Thomas.
C.D. Green (1995). 'All that glitters: a review of psychological research on the aesthetics of the golden section.' *Perception* 24: 937–68.

Our almost universal, probably inbuilt reactions to the faces of babies: C.B. Kennedy (1980). 'Attention of 4-month infants to discrepancy and babyishness.' *Journal of Experimental Child Psychology* 29: 189–201.

Babies evoke in us feelings of protectiveness: T.R. Alley (1983). 'Infantile head shape as an elicitor of adult protection.' *Merrill-Palmer Quarterly* 29: 411–27.

The features that make babies 'cuddly': J.B. Pittenger (1990). 'Body proportions as information for age and cuteness: Animals in illustrated children's books.' *Perception and Psychophysics* 48: 124–30.

Our responses to babies' faces and sales of commercial products: R.A. Hinde and L.A. Barden (1985). 'The evolution of the teddy bear.' *Animal Behaviour* 33: 1371–3.

Mickey Mouse's eyes were drawn in increasingly large proportions: S.J. Gould (1979). 'Mickey Mouse meets Konrad Lorenz.' *Natural History* 88: 30–6.

Babies with more pronounced infant faces are cuddled more: J.H. Langlois, J.M. Ritter, R.J. Casey and D.B. Sawin (1995). 'Infant attractiveness predicts maternal behavior and attitudes.' *Developmental Psychology* 31: 464–72.
J.M. Ritter, R.J. Casey and J.H. Langlois (1991). 'Adults' responses to infants varying in appearance of age and attractiveness.' *Child Development* 62: 68–82.

Certain people retain their babyish features into adulthood: L.A. Zebrowitz and J.M. Montepare (1992). 'Impressions of babyfaced males and females across the life span.' *Developmental Psychology* 28: 1143–52.

Babyfaced adolescents have their first sexual experiences later: A. Mazure, C. Halpern and J.R. Udry (1994). 'Dominant looking male teenagers copulate earlier.' *Ethology and Sociobiology* 15: 87–94.

Facial proportions of cover girls equivalent to a seven-year-old: D.M. Jones (1995). 'Sexual selection, physical attractiveness, and facial neotony.' *Current Anthropology* 36: 723–48.

The size of the baby's head key to releasing our protectiveness: D.S. Berry and L.Z. McArthur (1986). 'Perceiving character in faces: The impact of age-related craniofacial changes on social perception.' *Psychological Bulletin* 100: 3–18.

Research shows symmetrical faces more attractive: K. Grammer and R. Thornhill (1994). 'Human (Homo sapiens) facial attractiveness and sexual selection: The role of symmetry and averageness.' *Journal of Comparative Psychology* 108: 233–42. G. Rhodes, F. Proffitt, J.M. Grady and A. Sumich (1998). 'Facial symmetry and the perception of beauty.' *Psychonomic Bulletin and Review* 5 (4): 659–69

Infants stared significantly longer at faces the adults had rated most attractive: J.H. Langlois, J.M. Ritter, L.A. Roggman and L.S. Vaughn (1991) Facial diversity and infant preferences for attractive faces.' *Developmental Psychology* 27: 79–84.

Biology and symmetry in human facial attractiveness: R. Thornhill and K. Grammer (1999). 'The body and face of woman: One ornament that signals quality?' *Evolution and Human Behavior* 20 (2): 105–20.

Women are most symmetrical at the time they are most likely to conceive: J.T. Manning (1995). 'Fluctuating asymmetry and body weight in men and women: Implications for sexual selection.' *Ethnology and Sociobiology* 15: 145–53.

A high degree of symmetry may be an indication of particularly good genes: R. Thornhill and S.W. Gangestad (1993). 'Human facial beauty: Averageness, symmetry, and parasite resistence.' *Human Nature* 4: 237–69.

Birds and insects favour symmetry in their choice of mate: M. Ridley (1992). 'Swallows and scorpionflies find symmetry is beautiful.' *Science* 257: 327–8.

More symmetrical women had a greater number of sexual partners: R. Thornhill and S.W. Gangestad (1994). 'Human fluctuating asymmetry and sexual behavior.' *Psychological Science* 5: 297–302.

Symmetrical images of faces produced by doubling mirror images are not as attractive as we would expect: J.P. Swaddle and J.C. Cuthill (1995). 'Asymmetry and human facial attractiveness: Symmetry may not always be beautiful.' *Proceedings of the Royal Society of London* 261: 111–16.

Study shows attractive people are not necessarily healthier: S.M. Kalick, L.A. Zebrowitz, J.H. Langlois and R.M. Johnson (1998). 'Does human facial attractiveness honestly advertise health? Longitudinal data on an evolutionary question.' *Psychological Science* 9 (1): 8–13.

We find beautiful those faces which are more exaggerated: D.I. Perrett, K.A. May, S. Yoshikawa (1994). 'Facial shape and judgements of female

attractiveness.' *Nature* 368: 239–42.

Beautiful faces show more of those characteristics that we prefer: G. Rhodes and T. Tremewan (1996). 'Averageness, exaggeration, and facial attractiveness.' *Psychological Science* 7: 105–10.

Men and women choose their dates initially on physical attractiveness: A. Chapdelaine, M.J. Levesque, R.M. Cuadro (1999). 'Playing the dating game: Do we know whom others would like to date?' *Basic and Applied Social Psychology* 21 (2): 139–47.

A woman is more likely to be turned on to an attractive man when she is ovulating: J.C. Oliver-Rodriguez, Z. Guan and V.S. Johnston (1999). 'Gender differences in late positive components evoked by human faces.' *Psychophysiology* 36 (2): 176–85.

Research shows that women are attracted by a man's social and financial status: J.M. Townsend and G.D. Levy (1990). 'Effect of potential partners' physical attractiveness and socioeconomic status on sexuality and partner selection.' *Archives of Sexual Behavior* 19: 140–64.

Judgements of attractiveness based on images of faces morphed on a computer screen: D.I. Perrett, D.M. Burt, I.S. Penton-Voak, K.J. Lee, D.A. Rowland and R. Edwards (1999). 'Symmetry and human facial attractiveness.' *Evolution and Human Behavior* 20 (5): 295–307.

Women like men to look dominant, but not overly so: M.R. Cunningham, A.P. Barbee and C.L. Pike (1990). 'What do women want? Facialmetric assessment of multiple motives in the perception of male facial physical attractiveness.' *Journal of Personality and Social Psychology* 59: 61–72.

In 'lonely hearts' columns men put in information about their financial status: K. Deaux and R. Hanna (1984). Courtship in the personals column: the influence of gender and sexual orientation. *Sex Roles* 11: 363–75.

A large scale study shows that the most desired traits in a mate were kindness and intelligence: D.M. Buss (1994). *The Evolution of Desire: Strategies of Human Mating*. New York: Basic Books.

Warmth and kindness are the most important factors in mutual attraction: S. Sprecher (1998). 'Insiders' perspectives on reasons for attraction to a close other.' *Social Psychology Quarterly* 61 (4): 287–300.

The Wodaabe Tribe ceremony: C. Beckwith and A. Fisher (1999). *African Ceremonies*. Harry N. Abrams

In real-life choices of mate, attention is paid to personal qualities like kindness, intelligence and charm: R.E. Nisbett and T.D. Wilson (1977). 'Are the good beautiful or the beautiful good?' *Social Psychology Quarterly* 42: 386–92.

CHAPTER FIVE: VANITY

We know what a significant difference our appearance can make to the way others regard us: D.A. Santor and J. Walker (1999). 'Garnering the interest of others: Mediating the effects among physical attractiveness, self-worth and dominance.' *British Journal of Social Psychology* 38 (4): 461–77.

We try to present an image that reflects who we think we really are: D.A. Kenny and B.M DePaulo (1993). 'Do people know how others view them? An empirical and theoretical account.' *Psychological Bulletin* 114: 145–61.

Our ancestors chose to decorate their faces with red ochre: S. Mithen (1996). *The Prehistory of the Mind: The Cognitive Origins of Art, Religion, and Science*. London: Thames and Hudson.

The make-up kit of the Egyptian noblewoman was elaborate: R. Corson (1972). *Fashions in Makeup. From Ancient to Modern Times*. London: Peter Owen.

In Elizabethan times, women prized pale, translucent skin: J. Liggett (1974). *The Human Face*, chapters 3 and 4. New York: Stein & Day. Eighteenth-century French and British high society was outrageously extravagant: R. Corson (1991). *Fashions in Hair: The First Five Thousand Years*, p. 57. London: Peter Owen.

Studies show that women are usually considered more attractive when wearing make-up: T. Cash, K. Dawson, P. Davis, M. Bowen and C. Galumbeck (1989). 'Effects of cosmetics use on the physical attractiveness and body image of American college women.' *Journal of Social Psychology* 129: 349–55.

Age is the enemy of facial beauty for all of us: D.H. Enlow (1975). *Handbook of Facial Growth*. Philadelphia: Saunders.

A relatively stable appearance renders us predictable and safe: S.M. Kassin (1977). 'Physical continuity and trait inference: A test of Mischel's hypothesis.' *Personality and Social Psychology Bulletin* 3: 637–40.

On the whole, we are expected not to change ourselves: P.C. Bowman (1979). 'Physical constancy and trait attribution: Attenuation of the primacy effect.' *Personality and Social Psychology Bulletin* 5: 61–4.

'Persona' refers to the outward face of our psyche: C.G. Jung (1953). *Collected Works* (Bollingen Series). London: Routledge and Kegan Paul.

Piercing of facial areas has a long history: J. Liggett (1974). *The Human Face*, p. 46. London: Constable.

Today facial piercing is as common as personal decoration: D. McNeill (2000). *The Face*, p. 313. London: Penguin Books.

Tattooing the face has an ancient tradition: M. Mifflin (1997). *Bodies of Subversion: A Secret History of Women and Tattoo*. New York: Juno.

V. Vale (1998). *Modern Primitives: An Investigation of Contemporary Adornment and Ritual.* New York: Juno.

Many studies have shown that attractive people receive favourable treatment: K.K. Dion, E. Berscheid and E. Walster (1972). 'What is beautiful is good.' *Journal of Personality and Social Psychology* 24: 285–90.

A.H. Eagly, R.D. Ashmore, M.G. Makhijani and L.C. Longo (1991). 'What is beautiful is good, but...: A meta-analytic review of research on the physical attractiveness stereotype.' *Psychological Bulletin* 110: 109–28.

Teachers assume good-looking children to be more intelligent: M. Clifford and E. Walster (1973). 'Research note: The effects of physical attractiveness on teacher expectations.' *Sociology of Education* 46: 248–58.

Babies with characteristic 'cute' features receive more attention: K.A. Hildebrandt and H.E. Fitzgerald (1983). 'The infant's physical attractiveness: Its effect on bonding and attachment.' *Infant Mental Health Journal* 4: 3–12.

Children with disfigurements of the face attract the strongest rejection from their peers: K.H. Rubin and M. Wilkinson (1995). 'Peer rejection and social isolation in childhood: A conceptually inspired research agenda for children with craniofacial handicaps.' In R.A. Eder ed., *Developmental Perspectives on Craniofacial Problems*, pp. 158–76. New York: Springer-Verlag.

Attractive applicants receive higher starting salaries: R. Bull and N. Rumsey (1988). *The Social Psychology of Facial Appearance*, chapter 3. New York: Springer-Verlag.

Employment interviewers perceive attractive applicants as more competent: T. DeGroot and S.J. Motowidlo (1999). 'Why visual and vocal interview clues can affect interviewer's judgements and predict job performance.' *Journal of Applied Psychology* 84 (6): 986–93.

Plain people found guilty in court receive longer sentences than attractive people: A. DeSantis and W.A. Kayson (1997). 'Defendants' characteristics of attractiveness, race and sex and sentencing decisions.' *Psychological Reports* 81: 679–83.

M. Erian, C. Lin, N. Patel, A. Neal and R.E. Geiselman (1998). 'Juror verdicts as a function of victim and defendant attractiveness in sexual assault cases.' *American Journal of Forensic Psychology* 16 (3): 25–40.

Research has shown that people are generally more aggressive towards unattractive people: D. Alcock, J. Solano and W.A. Kayson (1998). 'How individuals' responses and attractiveness influence aggression.' *Psychological Reports* 82 (3): 1435–8.

Vanity leads to bullying and favouritism among schoolchildren: J.H. Langlois and L. Styczynski (1979). 'The effects of physical attractiveness on the behavioral attributions and peer preferences in acquainted children.' *International Journal of Behavioral Development* 2: 325–41.

Fat children are often shunned at school: G.R. Adams, M. Hicken and M. Salehi (1988). 'Socialization of the physical attractiveness stereotype: Parental expectations and verbal behaviors.' *International Journal of Psychology* 23: 137–49.

Winston Churchill's face: J. Brophy (1962). *The Human Face Reconsidered*, p. 186. London: Harrap.

Adolescents want to be seen with attractive peers: C.J. Boyatzis, P. Baloff and C. Durieux (1998). 'Effects of perceived attractiveness and academic success on early adolescent peer popularity.' *Journal of Genetic Psychology* 159 (3): 337–44.

Some people avoid individuals who have facial disfigurements: V. Houston and R. Bull (1994). 'Do people avoid sitting next to someone who is facially disfigured?' *European Journal of Social Psychology* 24: 279–84.

S.V. Stevenage and Y. McKay (1999). 'Model applicants: the effect of facial appearance on recruitment decisions.' *British Journal of Psychology* 90: 221–34.

CHAPTER SIX: FAME

The word 'icon' refers to portraits of spirit and divinity: D. McNeill (2000). *The Face*, p. 358. London: Penguin Books.

Elizabeth I portraits: R. Strong (1999). *The Cult of Elizabeth: Elizabethan Portraiture and Pageantry.* London: Pimlico.

Shakespeare's portrait. W.S. Pressly (1993). *A Catalogue of Paintings in the Folger Shakespeare Library.* New Haven: Yale University Press.

Faces of U.S. Presidents: C.F. Keating, D. Randall and T. Kendrick. (1999). 'Presidential physiognomies: Altered images, altered perceptions.' *Political Psychology* 20 (3): 593–610.

Tribal societies and the chief, warrior and shaman: B.C. Bates (1996). *The Wisdom of the Wyrd.* London: Rider/Random House.

So potent was the shaman's face, it could be represented as a mask: K. Johnstone (1981). *Impro: Improvisation and the Theatre*, p. 149. London: Methuen.

Shamanic rituals in Tibetan Buddhism: G. Samuel (1993). *Civilized Shamans: Buddhism in Tibetan Societies.* Smithsonian Institution.

Myths were gifts from the gods that could be understood only from within the imagination: S. Larsen (1998). *The Shaman's Doorway.* Inner Traditions.

The imagination is important to our emotional well-being: J. Hillman and M. Ventura (1993). *We've Had a Hundred Years of Psychotherapy and the World's Getting Worse.* San Francisco: Harper.

Pressman, T.E. (1992). 'The therapeutic potential of nonordinary states of consciousness.' *Journal of Humanistic Psychology* 32 (3): 8–27.

The word 'fan' means 'driven to a frenzy by worship of the divine': J.L. Caughey (1984). *Imaginary Social Worlds*, p. 40. University of Nebraska Press.

When the shaman was outlawed or displaced the actor took over: R. Taylor (1986). *The Death and Resurrection Show.* London: Anthony Blond.

Photography changed the nature of fame: M. Frizot, P. Albert and N. Harding, eds. (1999). *A New History of Photography.* Könemann.

Max Factor: F.E. Basten (1995). *Max Factor's Hollywood Glamour.* Los Angeles: General Publishing.

Identification, Positive and Negative Projection, and Hooks: J.R. Lewis, B.C. Bates and S. Lawrence (1994). 'Empirical studies of projection: A critical review.' *Human Relations* 47 (11): 1295–320

Commercials rely on our identifying with faces that almost always evoke a particular reaction: S. Brownlow and L.A. Zebrowitz (1990). 'Facial appearance, gender, and credibility in television commercials.' *Journal of Nonverbal Behaviour* 14: 51–60.

Mature faces more persuasive when selling a product based on their expertise: T. Onodera and M. Miura (1998). 'Interactive effects of physical attractiveness in advertisements'. *Psychological Reports* 83 (3): 1403–10.

Heroic characters in films are portrayed as good-looking: S.M. Smith, W.D. McIntosh and D.G. Bazzini (1999). 'Are the beautiful good in Hollywood? An investigation of the beauty and goodness sterotype on film.' *Basic and Applied Social Psychology* 21 (1): 69–80

We live simultaneously in the worlds of fantasy and reality: B.C. Bates (1986). *The Way of the Actor.* London: Century Hutchinson.

Film and television characters take on a 'reality': A. Evans and G.D. Wilson (1999). *Fame: The Psychology of Stardom.* London: Vision.

Cesare Lombroso tried to define the physical characteristics of evil: D. McNeill (2000). *The Face*, pp. 166–7. London: Penguin.

We identify, in our dreams and deep unconscious, with issues of spirit: M. Tucker (1992). *Dreaming with Open Eyes: The Shamanic Spirit In Contemporary Art and Culture.* London: Harper Collins.

Christopher Vogler and mythic structure in movies: S. Voytilla and C. Vogler (1999). *Myth and the Movies: Discovering the Mythic Structure in Fifty Unforgettable Films.* Michael Wiese Productions.

'Personality' movie actors: B.C. Bates (1986). *The Way of the Actor.* London: Century Hutchinson.

index

acknowledgements

The authors would like to thank, for their expertise and advice, everyone at BBC Worldwide, especially our editors Joanne Osborn and Sarah Lavelle, Linda Blakemore and Miriam Hyman; the BBC television production team, in particular Nick Rossiter, Sally George, James Erskine, Dave Stewart, Donald Sturrock, Amanda Lyon and Mary Harper; and also consultant Dr Sarah Stevenage for her advice on the text. We are grateful to all those whose personal stories were cited in the text, including Jim Cooke, Dan, Sharon and Lauren Deveney, Cindy Jackson, Vicky Lucas, Mirna Martines, the Needham family and Jennifer Thompson.

picture credits

BBC Worldwide would like to thank the following for providing photographs and for permission to reproduce copyright material. While every effort has been made to trace and acknowledge copyright holders, we would like to apologize should there be any errors or omissions.

The Advertising Archives 158, 213, 215; AKG London; 96 *top left*, 96 *centre left*, 96 *bottom left*, 170, 212; Brian and Cherry Alexander 184 *bottom left*; American Museum of Natural History 23; Andes Press Agency 240; Antoine Verglas, Inc. 2; Aquarius Picture Library 219, 220; Sven Arnstein 37 *left*; The Art Archive 167, 169, 210 *left*, *226*; BBC 99, 157 *right*; BBC/Bronwyn Kidd 13, 69, 86, 87, 112 *right*, 161, 195; BBC Natural History Unit 14 *bottom*, 16, 25 *top right*, 25 *bottom left*, 48, 114 *top right*, 142 *bottom*; BBC/ Richard Kendal 127; Linda Blakemore 64, 66, 67, 72, 80, 107; The Bridgeman Art Library 10, 24, 135 *top*, 139 *right*, 164, 166, 168, 169 *right*, 172 *above*, 180, 183, 199, 203; British Film Institute 196; Bruce Coleman 84 *top*, 84 *bottom*; Bubbles 29 *centre*, 49, 89 *bottom*, 102, 116, 141 *left*; Burlington Police Department, North Carolina 62; Camera Press 70 *right*, 95 *bottom right*, 96 *right*, 182 *bottom right*, 206; Capital Pictures 71, 132 *bottom left*, 132 *bottom*

right, 139 *left*, 157 *left*; Alex Cayley 41 *left*, Cindy Jackson Ltd/ www.cindyjackson.com 138 *left*, 138 *right*; Comstock 140; Corbis Images 36; Cramer Sweeney 55; Graham Davies 58; Sharon Deveney 76, 77; Professor Eibel-Eibesfeldt 89; Arther Elenaar and Remko Schaand 94, from 'Huge Harry': 'Towards a Digital Computer with a Human Face'. http://www.iaaa.nl, photographed by Josephine Jaspense; Sally and Richard Greenhill 141 *right*, 142 *top*, Professor Fumio Hara 124, 125; Hulson 79, 163, 189, 200; The Hutchinson Library 101, 209, 159 Image select 85; Images Colour Library 4, 35 *top*, 92 *left*, 92 *right*, 111 *top right*, 113, 173, *left*, 176, 184*top left*, 184 *bottom right*, 185, 208, 210 *right*, 223; Katz Pictures Ltd 128 *top left*, 128 *bottom right*, 130, 156 *right*, 190; The Kobal Collection 59; Magnum Photos 134; Museum of London 172 *bottom*, Natural History Museum 19; NHPA 111 *bottom*; Tony Notabaradino 41 *right*; Now/Vicky Lucas 192, courtesy of Changing Faces, 1/2 Junction Mews, London W2 1PN (tel. 020 7706 4232); Oxford Scientific Films 14 *top*, 111 *top left*, 225 *top left*; PA Photos 65 *left*, 65 *right*/by kind permission of the Needham family and South York-shire Police, England; Panos Pictures 29 *left*, 88 *top*; Performing Arts Library 188; Pictor International 27,

27, 30 *right*, 45, 46 *left*, 108; Pictorial Press 137; Picture Bank Photo Library 91, 230; Planet Earth Pictures 110; Popperfoto 204 *top*; Retna Pictures Ltd 2, 37 *left*, 37, *right*, *47*, 131, 151; Rex Features 46 *right*, 56, 70 *left*, 75 *left*, 75 *right*, 82, 105 *left*, 105 *right*, 128 *bottom left*, 129, 132 *top left*, 132 *top right*, 133, 135 *bottom*, 143, 144 *bottom*, 145, 146, 153, 154, 156 *left*, 174, 175, 177 *top*, 177 *bottom*, 178 *left*, 178 *centre*, 178 *right*, 179, 204 *bottom*, 216; Angela Fisher/Carol Beckwith, Robert Estall Photo Agency 38, 229; Robert Harding 100; The Ronald Grant Archive 144 *top*, 222, 224; Royal College of Surgeons of England 136; Royal Geographical Society 186; Science Photo Library 20, 26, 28, 78; Telegraph Colour Library 29 *right*, 33, 43 *left*, 52, 95 *top*, 114 *left*, 120, 148, 150, 155 *left*, 182 *top right*, 232; Tony Stone 7, 21, 30 *left*, 32, 34, 35 *bottom*, 42, 43 *right*, 44, 50, 54, 60, 73, 103, 112 *left*, 114 *bottom right*, 119, 122 *top*, 122 *bottom*, 147, 152, 162, 173 *right*, 181 *top left*, 181 *top right*, 181 *bottom right*, 182 *top left*, 182 *bottom left*, 184 *top right*; Topham Picturepoint 95 *bottom left*, 197, 198, 202, 218; Ugly Model Agency 155 *right*; Steve Watkins/BBC 207; The Wellcome Trust 31; Janine Wiedel 181 *bottom left*.